T0225192

Schools, Food and Social Learning

This book explores the potential of school dining halls as spaces of social learning through interactions between students and teachers.

Schools, Food and Social Learning highlights the neglect of school dining halls in sociological research and the fact that so much can be gained from fostering interpersonal relations with other students and the school staff over meals. The book focuses primarily on social and life skills that students develop during lunch-hour meetings, modelling behaviors while eating and conversing in the school space known as the 'restaurant'. With case studies based in the UK, the book takes a social constructivist approach to dealing with the tensions and challenges between the aims of the school – creating an eating space that promotes social values and encourages the development of social skills, and the activities of teachers and catering assistants of managing and providing food for many students daily.

The book carries snippets of interviews with children, dining hall attendants, teachers, parents and the school leadership team, offering a new way of thinking about social learning for both scholars and students of Social Anthropology, Sociology, Social Policy, Food Policy, Education Studies and Childhood Studies.

Gurpinder Singh Lalli is Senior Lecturer in Education and Inclusion Studies, based in the Institute of Education (IoE) at the University of Wolverhampton. Gurpinder has a vested interest in the sociology of education and is focused on school dining spaces. In 2017, he successfully defended a PhD in Education at the University of Leicester.

Schools, Food and Social Learning

Gurpinder Singh Lalli

LONDON AND NEW YORK

First published 2020
by Routledge
2 Park Square, Milton Park, Abingdon, Oxon OX14 4RN

and by Routledge
52 Vanderbilt Avenue, New York, NY 10017

Routledge is an imprint of the Taylor & Francis Group, an informa business

First issued in paperback 2021

British Library Cataloguing-in-Publication Data
A catalogue record for this book is available from the British Library

Library of Congress Cataloging-in-Publication Data
A catalog record has been requested for this book

ISBN: 978-1-138-36648-0 (hbk)
ISBN: 978-1-03-208851-8 (pbk)
ISBN: 978-0-429-43028-2 (ebk)

Typeset in Times New Roman
by Apex CoVantage, LLC

Contents

Tables

Acknowledgements

This book would not have come to fruition without the help and support of reviewers and close colleagues. I dedicate this book to my young family.

1 Introduction

Peartree Academy

It was a frosty winter's morning, at 8 o'clock on 16th January 2012 as I approached with some trepidation the school I later called Peartree Academy. The entrance was unusual. On the left, and built as part of the school, was a church. I discovered later that the church was regularly used by the local community, and ceremonies such as baptisms and marriages as well as funerals took place during the week even when the school was open. On the right was the school reception where all visitors were asked to sign in before proceeding. Walking straight ahead and through the doors into the school, I was faced with a surprise. The area that lay in front of me was a wide-open space filled with tables attractively grouped to provide seating for small to large numbers of people. On each table stood a small vase of fresh flowers and the whole eating area shone with cleanliness and care. Even though this was before the start of the school day, I saw children eating breakfast, parents talking to one another, and a few adults who I assumed to be the teachers chatting in groups, standing around the edge of the dining area. Later in the day, after I talked with the principal, I returned to the dining area, when children were coming in for lunch. This area was known as 'the restaurant' and it became my main research site for collecting data.

Background

It is often said that a 'family that eats together, stays together' (Stone, 2002: 270) and this book brings school food to the forefront as it leads to developing a new and refreshing way of thinking about school dining halls. The social glue in which this notion is based suggests that schools and families can be brought closer together in helping children to succeed in society and education, which influences their life chances. So what can be said about the ideal school dining space and what type of functions would it have? Is there any point in school dining hall reform or is it just a redundant part of the school? Whilst space for eating is accounted for in developed countries, it has not been so accessible for developing countries (McEwan, 2013; Jyoti et al., 2005; Ahmed, 2004); and as important as the calorific content of school lunch may be, the space in which it is being consumed is also salient and concentrated upon in this book. In addition, what type of learning opportunities can or could be presented in the school dining hall? In considering these questions, the book follows a narrative on how one particular school dining hall is operating and examples of the type of social learning that is taking place versus the tensions of whether forms of surveillance measures hinder this learning from occurring.

I have a vested interest and passion for school meals, not only to the extent of the calorific state of food but more so how food is being consumed and the social actors who take part in this activity. As a sociologist of education, I quickly became acquainted with the societal influences of inequality, yet my research has led me to research all things school food. Inevitably, this has involved speaking far and wide in conferences based in a number of subject disciplines including anthropology, education, geography, sociology and social policy. This multifaceted approach to school food has shaped me into a versatile thinker as one cannot ignore the multidisciplinary nature in which food exists, so this has connected me to a number of foodies both locally and internationally. Therefore, this book is particularly useful for sociologists, educators and leaders who have an interest in school dining spaces.

The sociology of food is a growing discipline which is yet to become cemented as a course programme in higher education, yet a number of programmes focus on food policy and the anthropology of food. Murcott (1983) discussed this emergence and highlighted how it would typically appear in newspaper articles. Whilst some literature is beginning to surface on this area, it is recent (Lalli, 2019; Earl, 2018; Wills et al., 2015; Andersen et al., 2015; Osowski et al., 2013); and Beardsworth and Keil (1996) also recognised the importance of the sociology of food and noted how the social sciences was just as important as the natural sciences in the study of human nutrition (Yudkin and McKenzie, 1964). Weaver-Hightower (2011) promotes the idea of placing a further emphasis on food-related research in an educational setting and recognises how food plays a vital role in the daily aspects of life in schools.

Policy

School meals are considered to be a popular area for policy reform; the work of the owners of restaurant chain LEON, Dimbleby and Vincent (DfE,

2013) introduced the *School Food Plan* (DfE, 2013) with the backing of the secretary of state for education at that time, Michael Gove, by devising a plan to support head teachers in school meal reform across UK schools. More recently, Henry Dimbleby was commissioned to lead an independent review to consider the food chain from field to fork, which will lead to the development of a National Food Strategy for England (DEFRA, 2019), reflective of a joint-up approach between government and industry experts. Dimbleby and Vincent (DfE, 2013) identify sustenance and the hidden benefits associated with food, such as how pupils, peers and teachers are able to sit and eat together whilst developing positive and lasting relationships in a civil environment. It was clear from the work of Dimbleby and Vincent (2013) that pupils cared about the food environment and like the idea of a clean and well-lit space, with friendly cooks and midday supervisors. It was also noted how pupils enjoyed socialising in the dining environment with peers, whether friends had packed lunches or school dinners, and the idea of a shorter queue was preferred. Close links to the School Food Plan (DfE, 2013) are discussed in a study by Hart (2016), placing a focus on the social context of the school meal, which highlights issues of public health and the school meal, a topic quickly becoming an area for public debate.

Historical perspective

The school meal is a symbolic and universal occasion, one which typically occurs at midday around the globe and this universal connection leads one to think about the influences upon meals which ultimately shape how they take place. For this reason, it is quite important to highlight a historical view of the English school meal in particular as there appears to be little work done on this area (Evans and Harper, 2009; Welshman, 1997) and the discussions to follow suggest that political agendas often feed into policy making. Three key periods in time have shaped the school meal to date which include 1) 1870–1879, 2) 1980–2000, and 3) 2001 to the present day.

1870–1979

The introduction of the Elementary Education Act (1870) set the framework for all children between the ages of 5 and 12 to start attending school. In 1880, this was made compulsory for children up to the age of 12. The driving force behind this was an apparent need for a more competitive Britain (Cross and MacDonald, 2009). Following the introduction of the Elementary Education Act (1870) in 1879, Manchester began to provide Free School Meals (FSM) to poorly nourished children, which saw the introduction of a similar scheme in Bradford, initially set up by Fred Jowett and

Margaret McMillan, who pushed for government legislation to encourage education authorities to provide school meals (Gillard, 2003). In a survey carried out between 1889 and 1903, it was found that a quarter of the population living in London did not have enough money to survive (Gillard, 2003). Moreover, Seebohm Rowntree's survey of working-class families in York in 1901 found almost half of those earning could not afford enough food to maintain physical efficiency (Gillard, 2003: 2). Consequently, due to levels of poverty at that time, children did not have access to appropriate nutrition and many parents did not understand nutrition due to the level of poverty (Gillard, 2003).

Following World War I, the introduction of the Education Act (1921) raised the school leaving age to 14 and empowered Local Education Authorities (LEAs) to provide FSM for those children who were eligible. However, due to the miners' strike of that year, attention was diverted, but eventually, the introduction of FSM led to an increase in the cost of providing meals, to almost £1m (Gillard, 2003). The Board of Education introduced a rationing system in order to limit the cost to central government, down to £300,000 (Welshman, 1997). Consequently, the rationing system affected the poor areas of the country, with less than half of those considered malnourished receiving meals (Webster, 1985: 216). Overall, a survey of 26 LEAs, carried out in 1936, showed that where unemployment was above 25 per cent, in a population of half a million, less than 15,000 children were receiving free meals, with eight of the LEAs having no service at all. By 1939, less than half of all local authorities were providing school meals, with 130,000 meals being served each day, totalling only 3 per cent of the school population, although 50 per cent were receiving milk (Smith, 1996: 191).

World War II saw a further emphasis placed on the nation's health, with food rationing introduced in 1940 as part of the war effort in an attempt to ensure a healthy nation. The school lunch had to be suitable as the main meal of the day and had to meet the nutritional standards (covering energy, protein and fat) introduced in 1941 (Morgan and Sonnino, 2010: 91). Eventually, school meals were introduced in all state schools during World War II. The Education Act (1944) made it compulsory for every LEA to provide a school meal, which became a significant feature of the welfare state (Gillard, 2003). LEAs were informed that the price of meals could not exceed the cost of food. During the Labour government (1945–1951), a proposal to provide all school meals free of charge was planned but eventually this was deemed unrealistic (Gillard, 2003). By 1951, 84 per cent of the population drank school milk. The typical daily diet of a child in 1951 included cereal or eggs with bread and butter for breakfast; meat, potatoes, a vegetable and a pudding for lunch; bread, butter, biscuits and jam for tea; with milk being the last thing at night (Gillard, 2003).

Up until the 1970s, the UK had a comprehensive school meals service, which was relatively cheap and also provided children from disadvantaged backgrounds the opportunity to access FSM. However, the introduction of the conservative government in 1979 meant the school meal had to adjust to the new consumer culture of the 1970s, which saw a change in attitude towards school meals. In her response to spending cuts, to meet their election pledges on tax, Margaret Thatcher demanded cuts in four areas, two of which included school meal charges and free school milk. When she became Prime Minister in 1979, during her first year, astonishingly for most, she brought an end to the provision of school milk for children over the age of seven (Smith, 2010). However, although she was known for abolishing free school milk, it was Harold Wilson's Labour government that stopped free milk for secondary pupils in 1968. In 1971, Thatcher, who was education secretary under Sir Edward Heath, brought an end to free school milk for children over the age of seven, although recent documents released suggest she had fought to save the grants but was overruled by Sir Edward Heath (Smith, 2010).

1980–2000

The Education Act (1980) also saw a move away from meeting nutritional standards, which involved an end to free school milk whilst removing any obligation for meals to be sold at a fixed price. This enabled LEAs to provide free milk as the scheme enabled them to claim additional funding for primary and secondary milk sales. In 1988, many children lost their eligibility for FSM and some school meal payments made by the government were replaced by direct cash sums to families. However, there was no way of establishing whether this cash incentive was spent on food. By 1990, the criteria for FSM changed as LEAs were only required to provide meals for children entitled to free dinners and also to provide a place for children to eat packed lunches (Dare and O'Donovan, 2002: 90). This saw the introduction of a privatised culture of school meals, where the priority shifted from providing a nutritional meal with in-house school control to a 'value for money' meal culture.

By the time New Labour took power in 1997, there was a mass of evidence pointing to the nation's health concerns, particularly children's diets, which had become less healthy over time, with concerns about excessive levels of sugar, salt and fat (Gillard, 2003). The government announced that it would introduce nutritional guidelines which encouraged school canteens to provide a choice of four main categories of food – fruit and vegetables, meat and protein, starchy foods and milk and dairy products. The main agenda was to ensure fruit and vegetables were accessible and affordable for all, as the Public Health Minister, Yvette Cooper, argued in 2000 that children who grew up in low income households ate less fruit and vegetables than children who grew up in high income households.

2001–present day

In 2001, over 1.8 million children in the UK were eligible for FSM, but it was reported that only one in five pupils claimed this entitlement (Storey and Chamberlain, 2001). Nutritional standards were reintroduced in 2001. It was during the period 2001–2011 that attention was refocused on health matters, and in particular, issues of obesity and quality of school dinners. These concerns were pushed into political focus in February 2005 by food chef Jamie Oliver, in the television series *Jamie's School Dinners*. For once, this brought a general consensus amongst the three main government parties (Labour, Conservatives and the Liberal Democrats). Their manifestos leading up to the general election all promised to improve school meals, police junk food advertising aimed at children and to control the content of school vending machine sales (Gillard, 2003). Education Secretary Ruth Kelly promised an additional £280m to improve school meals, and as of March 2005 the government was required to allocate 50 pence a day on ingredients per primary pupil and 60 pence per day per secondary pupil (Branigan et al., 2005).

In 2005, food revolutionary Jamie Oliver's Channel 4 series *Jamie's School Dinners* (Conlan, 2005) caused an uproar because evidence identified pupils consuming a quarter of a ton of chips every week at Kidbrooke School in Greenwich (BBC News, 2007). The food budget was 37p per meal and the dinner ladies at the school had become demotivated. As a result, the Children's Food Trust (CFT) was set up in 2005 in order to help schools introduce and maintain the national standards for school food, as well as helping children enjoy their lunchtime experience. Junk food was banned in schools in 2006 (BBC News, 2006).

In 2007, OfSTED introduced new interim standards for food in schools in 2006. A survey evaluated the progress schools were making in meeting the new standards, which would ensure that school lunches provide pupils with a healthy diet (OfSTED, 2007: 7). However, the report identified that school lunch take-up had fallen in 19 of the 27 schools visited for the survey (Curtis, 2007). The reasons for the decline included a lack of consultation with pupils and parents about the new arrangements in school, poor marketing of new menus, high costs for low income families who were not eligible for FSM and also a lack of meal choice (OfSTED, 2007: 5). In addition, dining areas played a part in the take-up of school meals as they varied in quality, with primary school generally being better than secondary school dining spaces. Overall, pupils wanted shorter queuing times in order to allow time for extra-curricular activities and the opportunity to eat with friends who brought packed lunches.

In April 2012, as mentioned earlier, former education secretary Michael Gove requested the services of John Vincent and Henry Dimbleby (founder of LEON restaurants) to carry out an independent review of school food (Long, 2018). This led to the publication of the School Food Plan (DfE, 2013). Recommendations from the School Food Plan (DfE, 2013: 10) were targeted at head teachers, as those who could influence the vision of schools in adopting a more forward-thinking mentality. Littlecott et al. (2015) provided evidence to suggest that children who eat a healthy breakfast before starting the school day achieve higher academic results than pupils who do not. This study was carried out using a sample of 5,000 students aged 9–11 from more than 100 primary schools in England. Ten years on from the exposure of *Jamie's School Dinners*, he admitted that his campaign was far from a success based on the notion that eating well is still viewed as an indulgence of the middle class (Furness, 2015).

School food programmes

It was particularly interesting to review School Food Programmes (SFPs) across the globe as I was able to discover key issues specifically relating to participation and the influence of food on pupil engagement in school. SFPs were introduced in developing countries to provide nutritional meals to reduce short-term hunger in the classroom and increase the attention span of pupils (Ahmed, 2004). Furthermore, Kleinman et al. (2002) carried out a study to investigate academic performance in the United States and found that 33 per cent of children who were classed as being at nutritional risk had significantly poorer test scores, attendance and punctuality, compared with their counterparts who were not at risk.

Academic performance

There are a multitude of variables involved in the academic development of a child (Lerner and Jovanovic, 1999). Such variables could include the social space in which a child learns how to behave and develop the relevant skills to perfect the social rules that govern a particular site. Jyoti et al. (2005) explored the links between food insecurity, academic development and social skills for both male and female pupils. This study used longitudinal data, which highlighted financial barriers leading to poor nutrition and the consequences for academic performance and social skills for pupils. Overall, the study found strong evidence arguing food insecurity is linked to non-nutritional issues such as weak academic performance for boys and girls. Girls in particular were identified as suffering from poor social skills and low reading abilities. This led me to question the difficulty in trying to develop a discussion between school and academic performance, and by academic performance I am referring to the potential informal learning opportunities presented in the school dining hall, later introduced as social learning. Moreover, the type of learning that is explored is one which takes place away from the classroom environment.

The links to nutritional status in developing countries are inevitable as the nourishment and participation of children in consuming school food is a key issue. For instance, a number of past studies were identified through the works of Ahmed (2004), conducted in Chile and Jamaica (Pollit, 1990; Simeon and Grantham-McGregor, 1989), that addressed nutritional status and academic performance in relation to SFPs. It is said that children who do not perform academically are often subject to poor nutritional status (Pollit, 1990). According to findings from the Food Research and Action Centre (FRAC, 2011), missing breakfast impaired a child's ability to learn, whilst consuming breakfast improved their academic performance and behaviour. FRAC (2011) noted that those who missed breakfast were less able to differentiate between visual images with a slower memory recall resulting in errors. FRAC (2011) and Murphy (2007) also reported that children who ate a complete breakfast made fewer mistakes and worked much quicker in the classroom arithmetic assessments. Moreover, Murphy (2007), who conducted a study in the United States, found that in the short term, breakfast skippers had less energy available and were undernourished over sustained periods. For Murphy (2007), these children were more likely to feel hungry and less likely to be active.

Whilst there is no clear evidence to suggest that school food has a direct link to academic performance, numerous studies based in developing countries explore this field of inquiry, but the concluding ideas point to the notion that food alone is not the answer to improving nutritional status. For this reason, the emphasis on the social aspect of school food is heightened. For example, the findings from a study carried out in Bangladesh to investigate the outcomes of the SFPs pointed to the lack of participation, academic achievement in primary education in developing countries and identified how there were two causes for this problem (Ahmed, 2004). These causes included a lack of health and nutrition, which ultimately affected the pupils' ability to learn. A number of studies conducted in Ethiopia and a number of other developing countries (Clay and Stokke, 2000; Pelletier et al., 1995) highlight how food alone does not guarantee improved nutritional status (Ahmed, 2004). Furthermore, if nutritional status is not guaranteed, this has an impact on the pupils' learning and attention span in the classroom

(Ahmed, 2004). The study carried out in Ethiopia (Pelletier et al., 1995) highlighted other reasons apart from access to food that had an impact on children's nutritional status. The other reasons addressed child caring, feeding practices and household variables such as income and parental education. In developing countries poor health facilities and services were also barriers in a child's nutritional wellbeing (Pelletier et al., 1995).

Whilst the research in developing countries focuses on the nourishment of children in participating and consuming school food whilst trying to inform academic performance, the discussion in developed countries also holds a firm grip on linking food to academic performance and nutrition, but works on the assumption that there is already an established level of nourishment. This means that other aspects that include the 'social' can also be considered to improve the mealtime experiences of children.

Behaviour, socialisation and promoting health

Similar to the issues discussed in the developing countries, it is evident that the issues are alike in developed countries in terms of pupil participation in school meal up-take. For example, data from a research project carried out in four Finnish schools on school diet preferences and behaviour found that pupils receive one quarter of their daily energy from school meals, which highlights the potential influences of those who are served meals to their ongoing diet and attitude towards eating (Tikkanen, 2009). One of the reported issues in this project is that pupils were not always keen on eating at school, whether meals were free or run on a commercial basis (Tikkanen, 2009). Although school meals are free of charge in Finland and Sweden, some pupils still choose not to eat certain parts of the meal. Meal choices certainly appear to be an important factor with regards to the school meal experience that pupils have when consuming food in the dining area. Whilst the issue of school meal participation can be comparable to the one discussed in developing countries, this appears to be about the selection and options available in terms of food consumption.

Overall, the shortage of research carried out on SFPs suffice to say that there are multiple conclusions to be taken away from this for my future research. The majority of studies based in developing countries pays close attention to health and wellbeing and make an attempt to demonstrate a link between school food and academic performance. The same can be said for developed countries, although whilst access to nourishment is not comparable to the lack of access to food available in developing countries, the focus on health and academic performance is also evident. The provision of the way in which food can help children in other ways away from the classroom environment are only partially discussed in the developing world, however, compared to the developed nations which seem to have a growing trend towards highlighting the opportunities of how school food can help to foster social learning.

Context

Peartree Academy is an all through 3–16 urban school which was rebuilt in 2007. The school replaces what was a deteriorating school in a deprived area of a UK City. According to the latest figures (DfE, 2019) it holds a capacity for 1046 pupils with 933 currently enrolled and 272 staff. The school specialises in business and enterprise with a focus on food. Local businesses also are involved within the school, by supplying the ingredients for the school breakfast club as well as sending in their own chefs to teach cookery skills.

It is a mixed-sex school with a large proportion of pupils of White British heritage. The proportion of pupils supported by the pupil premium is above average. The pupil premium is an initiative which allows access to additional government funding in order to support disadvantaged pupils, who are identified as being eligible for (FSM. A total of 51 per cent of pupils at Peartree Academy have access to FSM. In addition to this, 49 per cent of pupils have Special Educational Needs (SEN), which is above average for a school. According to the 2011 census, income, employment, health and education deprivation was recorded as high, in comparison to the rest of the UK. The unemployment rate was measured at 13.3 per cent, which was 7.8 per cent across the UK. It is useful to bear these figures in mind when interpreting the contextual demographics of the school.

The school curriculum is made up of three phases. Phase 1 includes the Early Years Foundation Stage (EYFS) which runs through to Year 4, with a particular focus on literacy, numeracy, personal and social skills in preparation for the wider curriculum. Phase 2 includes Years 5–8 which continue to develop literacy and numeracy skills in preparation for the GCSE stage. The school's curriculum in Phase 3 supports Years 9–11 in making subject choices alongside the school's core curriculum of English, Maths, Science, RE (Religious Education), PE (Physical Education) and Enterprise. The school is innovative in its role in supporting the local community, through raising standards for pupils as well as providing a strong support network for parents and the wider community and giving them access to developing their skills further to aid local regeneration. For example, unpaid and paid employment is available, which has seen some parents working as teaching assistants in the classroom and others in the canteen as assistant chefs and lunchtime supervisors.

The principal's philosophy of the school rests on trying to exceed set expectations each year in order to support children in becoming successful adults. For the principal, the restaurant acts as the focal point of the school.

Staff members at the school are able to use the restaurant to build positive relationships with parents whilst discussing children's progression. In addition, parents and community members are able to take up opportunities to volunteer as support staff in the school, both in and outside of the classroom.

The board of governors within the school plays an influential role in the day-to-day running of the school. It includes two parent governors, one of whom is responsible for the extended school activities. This is an opportunity for the local community to develop a wide range of skills in sports and arts and gain further prospects towards achieving additional GCSEs. A local businessman who funded the school has been involved in the food manufacturing business and is very passionate about the importance of having a strong team of governors, which includes education professionals to be able to drive the ethos of the school forward.

Peartree Academy is a Church of England (CoE) school and there is a strong impetus towards helping the pupils to achieve their full potential throughout their early and adolescent years. The school is designed with the church positioned on the left as a key feature so that everyone who enters the school will pass the church. The Church at the school is closely linked to the community, which enables pupils to understand the meaning of belonging in a community. The Christian values of the school are represented through the school's core values, which include loving, caring and respecting, forgiveness and new beginnings, trust and honesty, nurturing, faith and prayer and 'doing our best'.

Research questions

In January 2012, I remember reading about childhood obesity and school lunches in a newspaper which was obtained on route to a connecting flight from the airport during a family vacation to Edmonton, Canada. This was my initial connection to my research; I had been given the opportunity to begin developing a research proposal in October 2011. Prior to starting the writing of this research project, I noticed how most studies were carried out in an international context (McEwan, 2013; Jyoti et al., 2005; Ahmed, 2004), which led to using certain search criteria and formulating a set of research aims which appeared in the form of three questions, with one overarching question followed by two subsidiary questions. These questions, presented next, formed the basis of my research investigation.

1   What is the impact of the food environment upon social learning?
2   How do eating behaviours of staff and pupils impact on social learning?
3   How do teaching staff promote social learning within a food environment?

In terms of collecting the data for this work, I used an ethnographic (Hammersley and Atkinson, 2007) approach which involved embracing myself as a participant in the everyday life of the school in which interviews were recorded, field notes were taken and observations were noted in exploring the interactions that take place in one school dining hall.

References

Ahmed, A. U. (2004) *Impact of Feeding Children in School: Evidence from Bangladesh*. Washington, DC: International Food Policy Research Institute.

Andersen, S. S., Holm, L. and Baarts, C. (2015) School meal sociality or lunch pack individualism? Using an intervention study to compare the social impacts of school meals and packed lunches from home, *Social Science Information*, pp. 1–23.

BBC News (2006) 'Junk food banned in school meals', 19 May. Available at: http://news.bbc.co.uk/1/hi/education/4995268.stm (Accessed: 10 August 2019).

BBC News (2007) 'Jamie's top dinner lady to quit', 3 April. Available at: http://news.bbc.co.uk/1/hi/education/6521213.stm (Accessed: October 2019).

Beardsworth, A. and Keil, T. (1996) *Sociology on the Menu: An Invitation to the Study of Food and Society*. London: Routledge.

Branigan, T., Lawrence, F. and Taylor, M. (2005) 'Kelly passes school dinner test', *The Guardian*. Available at: www.theguardian.com/society/2005/mar/31/schools.politics2 (accessed: 10 March 2019).

Clay, E. and Stokke, O. (2000) *Food Aid and Human Security Book*. London: Frank Cass.

Conlan, T. (2005) 'Channel 4 asks for more school dinners', *The Telegraph*. Available at: www.theguardian.com/media/2005/aug/19/schools.channel4 (accessed: 12 March 2019).

Cross, M. and MacDonald, B. (2009) *Nutrition in Institutions*. London: Wiley-Blackwell.

Curtis, P. (2007) 'Children find Jamie Oliver's school food hard to swallow, say inspectors', *The Guardian*, 3 October. Available at: https://www.theguardian.com/society/2007/oct/03/2 (Accessed: 4 September 2019).

Dare, A. and O'Donovan, M. (2002) *A Practical Guide to Child Nutrition*. Cheltenham: Nelson Thornes.

Department for Education. (2013) 'The school food plan: How to improve school food and schoolchildren's diets'. Available at: www.gov.uk/government/publications/the-school-food-plan (accessed: 10 July 2019).

Department for Education. (2019) 'Peartree Academy'. Available at: www.education.gov.uk/cgi-bin/schools/performance (accessed: 11 June 2019).

Department for Environment, Food and Rural Affairs. (2019) 'Developing a national food strategy: Independent review'. Available at: www.gov.uk/government/publications/developing-a-national-food-strategy-independent-review-2019 (accessed: 4 July 2019).

Dimbleby, H. and Vincent, J. (2013) 'The school food plan, Department for Education'. Available at: www.schoolfoodplan.com/wp-content/uploads/2013/07/School-Food-Plan-2013.pdf (accessed: 10 August 2019).

Earl, L. (2018) *Schools and Food Education in the 21st Century*. London: Routledge.

Education Act. (1921) Available at: www.educationengland.org.uk/documents/acts/1921-education-act.html (accessed: 2 July 2019).

Education Act. (1944) Available at: www.legislation.gov.uk/ukpga/Geo6/7-8/31/contents/enacted (accessed: 8 February 2019).

Education Act. (1980) Available at: www.legislation.gov.uk/ukpga/1981/60/enacted (accessed: 10 February 2019).

Elementary Education Act. (1870) 'The 1870 Education Act'. Available at: www.parliament.uk/about/living-heritage/transformingsociety/livinglearning/school/overview/1870educationact/ (accessed: 20 February 2019).

Evans, C. E. L. and Harper, C. E. (2009) 'A history and review of school meal standards in the UK', *Journal of Human Nutrition and Dietetics*, 22, pp. 89–99.

Food Research and Action Centre (FRAC). (2011) 'Breakfast for learning: Scientific research on the link between children's nutrition and academic performance', 1–4. Available at: http://frac.org/wp-content/uploads/2011/08/breakfastforhealth.pdf (accessed: 18 February 2019).

Furness, H. (2015) 'Jamie Oliver admits school dinners campaign failed because eating well is a middle class preserve', *The Telegraph*. Available at: www.telegraph.co.uk/news/celebritynews/11821747/Jamie-Oliver-admits-school-dinners-campaign-failed-because-eating-well-is-a-middle-class-preserve.html (accessed: 23 June 2019).

Gillard, D. (2003) 'Food for thought: Child nutrition, the school dinner and the food industry', *Forum*, 45 (3), pp. 111–118.

Hammersley, M. and Atkinson, P. (2007) *Ethnography: Principles in Practice*. London: Routledge.

Hart, C. S. (2016) 'The school food plan and the social context of food in schools', *Cambridge Journal of Education*, 46 (2), pp. 211–231. Available at: http://eprints.whiterose.ac.uk/96751/22/5-18-2016_The%20School.pdf (accessed: 10 January 2019).

Jyoti, D. F., Frongillo, E. A. and Jones, S. J. (2005) 'Food insecurity affects school children's academic performance, weight gain and social skills', *American Society for Nutrition*, pp. 2831–2839.

Kleinman, R. E., Hall, S., Green, H., Korcez-Ramirez, D., Patton, K., Pagano, M. E. and Murphy, J. M. (2002) 'Diet, breakfast and academic performance in children', *Annals of Nutrition Metabolism*, 46 (1), pp. 24–30.

Lalli, G. (2019) 'School meal time and social learning in England', *Cambridge Journal of Education*. DOI: www.tandfonline.com/eprint/fHGEr7k8ebj55DVBY6YN/full?target=10.1080/0305764X.2019.1630367.

Lerner, R. M. and Jovanovic, K. (1999) *Cognitive and Moral Development, Academic Achievement in Adolescence*. London: Routledge.

Littlecott, H. J., Moore, G. F., Moore, L., Lyons, R. A. and Murphy, S. (2015) 'Associations between breakfast consumption and educational outcomes in 9–11 year old children', *Public Health Nutrition Journal*, 19 (9), pp. 1–8.

Long, R. (2018) 'School meals and nutritional standards: Briefing paper', House of Commons: Parliament [Online] Available at: http://dera.ioe.ac.uk/31079/1/SN04195%20_Redacted

McEwan, P. J. (2013) 'The impact of Chile's school feeding program on education outcomes', *Economics of Education Review*, 32, pp. 122–139.

Morgan, K. and Sonnino, R. (2010) The Urban Foodscape: World Cities and New Food Equation, *Cambridge Journal of Regions, Economy and Society*, Oxford University Press.

Murcott, A. (ed.) (1983) *The Sociology of Food and Eating*. Aldershot: Gower.

OfSTED, Office for Standards in Education. (2007) *Food in Schools: Encouraging Healthier Eating*. London: OfSTED. Available at: http://dera.ioe.ac.uk/1124/1/Food%20in%20schools.pdf (accessed: 20 February 2019).

Osowski, C. P., Goranzon, H. and Fjellstrom, C. (2013) 'Teachers' interaction with children in the school meal situation: The example of pedagogic meals in Sweden', *Journal of Nutrition Education and Behaviour*, 45 (5), pp. 420–427.

Pelletier, D. L., Denee, K., Kidane, Y., Haile, B. and Negussie, F. (1995) 'The food-first bias and nutrition policy: Lessons from Ethiopia', *Food Policy*, 20 (4), pp. 279–298.

Pollit, E. (1990) 'Malnutrition and infection in the classroom: Summary and conclusions', *Food and Nutrition Bulletin*, 12 (3).

Simeon, D. T. and Grantham-McGregor, S. (1989) 'Effects of missing breakfast on the cognitive functions of school children of differing nutritional status', *American Journal of Clinical Nutrition*, 49 (4), pp. 646–653.

Smith, D. (1996) *Nutrition in Britain: Science, Scientists and Politics in the Twentieth Century*. London: Routledge.

Smith, R. (2010) 'How Margaret Thatcher became known as "Milk Snatcher"', *The Telegraph*. Available at: www.telegraph.co.uk/news/politics/7932963/How-Margaret-Thatcher-became-known-as-Milk-Snatcher.html (accessed: 28 February 2019).

Stone, L. (ed.) (2002) *New Directions in Anthropological Kinship*. Lanham, MD: Rowman and Maryland.

Storey, P. and Chamberlain, R. (2001) 'Improving the take up of free school meals', in *Research Brief 270*. London: DfE.

Tikkanen, I. (2009) 'Pupils' school meal diet behaviour in Finland: Two clusters', *British Food Journal*, 111 (3), pp. 223–234.

Weaver-Hightower, M. B. (2011) 'Why education researchers should take food seriously', *Educational Researcher*, 40 (1), pp. 15–21.

Webster, C. (1985) 'Health, welfare and unemployment during the depression', *Past and Present*, 109 (1), pp. 204–230.

Welshman, J. (1997) 'School meals and milk in England and Wales, 1906–45', *Medical History*, 41, pp. 6–29.

Wills, W., Draper, A. and Gustafsson, U. (2015) *Food and public health: Contemporary issues and future directions*. London: Routledge.

Yudkin, J. and McKenzie, J. (1964) *Changing Food Habits*. London: McGibbon and Kee.

2 Social learning

Introduction

Having provided a discussion on multiple issues within which the school meal is bounded, it is important to introduce a discussion on social learning. This chapter is made up of four parts which include 1) changes to the school environment, 2) social learning, 3) the pedagogic meal, and 4) food education and intervention. Ultimately, the work here set out to explore the social aspect of the school meal at Peartree Academy. This led me to review a number of relevant studies which share the view that the school meal should be prioritised and more importantly should be integrated into school life and learning. A number of studies (Lalli, 2019; Hart, 2016; Osowski et al., 2013; Pike and Leahy, 2012; Delormier et al., 2009) explore the social context of the school meal in connection with diet, behaviour and food. Other studies (Sepp et al., 2006; Fjellstrom, 2004) introduce the notion of the 'pedagogic meal'; a way the food situation can be used as a tool to promote informal education. This is relevant to my research as I also explore the link between food and learning. Durlak and Weissberg (2007) address how social skills are being developed through certain programmes in the United States, specifically extra-curricular activities, whilst the work of Benn and Carlsson (2014) identifies how learning opportunities are developed in the school dining hall by means of intervention.

Changes to the school environment

There is much to be said about changes to the school environment in terms of the facilities and it is worth exploring this in relation to the school restaurant at Peartree Academy as a context. These were some of the questions I had asked myself and this led to reviewing studies which explored environmental changes to school dining halls. A number of studies highlight the impact of the school environment in encapsulating and encouraging learning opportunities to take place (Woolner et al., 2018; Garner, 2011; Devi et al., 2010; Banerjee, 2010). Rudd et al. (2008) carried out 'before' and 'after' surveys in a UK high school with two year groups to see whether their attitudes to school and learning had changed following a revamp to the school. They found a strong association between students' attitudes with the move towards improved facilities. The survey also identified with students' outlooks on the future as being positive. Moreover, students were questioned about 'spaces' and 'places' within the school, where they felt they learned most whilst enjoying activities with peers. Students were asked before and after the changes. Before the survey, social spaces in and around the school made up 37 per cent of responses and after the survey and revamp made up 43 per cent. Students were also questioned on the overall school facilities in order of importance. Sixty-four per cent of students identified 'good dining facilities for healthy eating' as very important.

Therefore, the school dining hall is one space which can be used to facilitate change and have an impact on behaviour and learning opportunities. An investigation was carried out in a middle school to identify how changes to the school environment could be developed to improve physical activity and nutrition (Bauer et al., 2004). They identified time constraints as a barrier in eating nutritious food. Students pointed out that once they had finished queuing up and collecting their food, they did not have the time to eat, which resulted in them often choosing an unhealthy option from the snack cart – something they were able to consume quickly (Bauer et al., 2004: 41). Overall, Bauer et al. (2004) found numerous pressures across school environments, which need to be examined in order to help develop ways of improving tsocial and physical activity, including nutrition. Therefore, the impact of a poorly designed dining hall can have a negative impact on areas such as food choices for children and the reduced time for the lunches due to longer queues can also have a negative impact.

It is said that the school meal is able to present opportunities for changing the behaviours and attitudes of children in schools, through a 'whole-school approach' (DuCharme and Gullotta, 2012). A whole-school approach is about the design and co-ordination of school standards in considering pupil-staff roles across settings in relation to the development needs of pupils (DuCharme and Gullotta, 2012). McNeely et al. (2002) examined the association between school connectedness and the school environment to identify ways to keep students more attached to the school. The study was conducted as longitudinal research, where 80 schools were selected at random in order to measure the level of the students' connectedness to the school. McNeely et al. (2002) identified the importance of schools as places for intervention in terms of supporting student health, which include nutrition programmes and PE in the curriculum. However, they also supported the idea that although these programmes were effective and important in their own right, they did not address the crucial requirement of student health, which is their wellbeing (McNeely et al., 2002: 145). It is the feeling of belonging that is missing and the feeling of being cared for through pastoral pursuits. The whole-school approach adopted by Peartree Academy prioritises the food environment in promoting a healthy living.

It was identified that establishing and maintaining a safe and calm school environment was instrumental in helping to facilitate pupils' learning and opportunities, as well as promoting opportunities for progression (Banerjee, 2010). This study also highlights the importance of the school conditions in which pupils are socialised. There are certain characteristics of how the environment has been built and set up at Peartree Academy. There are a number of factors which influence pupil behaviour, particularly the wellbeing of pupils, which is often affected by broader issues including the overall school environment (Banerjee, 2010: 7). Banerjee (2010) also pointed out how a 'whole-school universal' approach was the key to addressing issues of pupil wellbeing and promoting the idea of creating a connectedness amongst pupils as well as a safe and calm learning environment, further promoting learning, positive behaviour and positive peer interactions. Weir (2008) also supported the notion of taking a whole-school approach, placing particular emphasis on the standards of school meals. These views were taken from a caterer's experiences and perceptions of the school meal. Weir (2008) identified how caterers argued that all those involved in schools needed to take responsibility in order to provide support through the ethos of a whole-school approach.

Furthermore, in terms of environmental factors, school dining halls are commonly associated with having issues which create certain barriers. For example, Devi et al. (2010) explored the environmental factors influencing schools' decisions and children's food choices in relation to vending machines. It was found that structural factors in the school canteen were the second most problematic of issues concerning school meals (Devi et al., 2010: 217). More specifically, this included canteen facilities, increased numbers of pupils entering the canteen, shortened lunch breaks and disruption of long queues. From the focus group interviews, students identified their frustrations in making food choices; these were due to the long queues, short lunch breaks, seating space available and unattractiveness of the school canteen (Devi et al., 2010: 218). This provides further evidence of the day-to-day practical problems with the school environment associated with the school dining hall.

Having discussed the issues surrounding the school meal, it is useful to consider the views of children in establishing the most suitable type of school dining space. Dudek (2005) introduced the idea of an edible school space, suggesting that it involved drawing on the views of children about their ideal school environments. The study identified how the school meal as an edible landscape painted an impression of the school. He also reported how the ideal school spaces devoted to the preparation and consumption of food should be detached from the rest of the school building. In the UK, some schools tend to make multiple use of the school dining area, including use as an assembly hall and space for PE lessons. Dudek (2005) discusses how the school hall has traditionally been known as a space for dining which has been experienced by pupils and teachers over generations (Dudek, 2005: 260). He then distinguishes between the workplace for adults where food consumption takes place in a separate area detached from production or manufacturing spaces. The key point here is that the priority for the area in which school food is served has not been the primary concern for schools.

Symbolic attachments to school buildings may bring about potential benefits to the school meal experience. In a study by Nicholson (2005) the school building was identified as the third teacher. Following project work in communities based in California, she identified the importance of the environmental factors and in this case the school building as a tool for fulfilling ideas about how children learn, what they learn and how they are taught (Nicholson, 2005: 45). Aside from educational objectives, Nicholson (2005) discussed how the school building acted as a pillar for building respect in children, whilst making a difference in their life experiences. She discussed how every aspect of an educational environment represented a choice regarding the provision to be offered and identified three key struggles in school life: 1) the struggle for time, 2) use of space, and 3) use of money. For Nicholson (2005) even choices made by the school were seen to carry certain values and symbolic messages. This highlights the importance of the school environment and perhaps the reason why the design of school dining halls have been influenced.

Social learning

With regards to finding a suitable definition, I set out to explore multiple views of how social skills had been interpreted and also decided to include the term 'learning' in my search criteria, in order to come up with a holistic definition that I could then carry forward for my research. For Dalton (2004), social skills are needed by all members of society; I would like to unpack this by stripping the words down individually. 'Social' can be defined as society, concerned with mutual relations of human beings, living in organised communities'; and 'skill' can be defined as 'expertness; practised ability' (Dalton, 2004: 14). For the purpose of this study, I am defining 'social' as being part of an organised community (activity in the restaurant) and 'learning' as including skills which have to be practiced (i.e. using a knife and fork) as well as knowing how to behave in a social context (i.e. good manners, politeness and modelling behaviour). Whilst this is the definition that is carried forward throughout my study, it is useful to review other definitions in order to look closely at how these terms have been interpreted.

Some of the literature on social skills in a school-based context has been conducted around Asperger's syndrome as the three main areas of difficulty as listed on the diagnosis, which include 1) social interaction, 2) social communication and 3) social imagination (DoH, 2014). As behaviour is a key part of how interactions are taking place in the school restaurant it is important to draw on some of the views of commentators and their definitions of what constitute 'social skills' (Bellack et al., 2004; Dalton, 2004).

It is argued that there is no single definition of what constitutes a social skill and a situation-specific conception of social skills would be more relevant. Bellack et al. (2004) recognised the neglect in connection to the broad array of social behaviours in considering a suitable definition of social skills. The overriding factor is the effectiveness of behaviour in social interactions (Bellack et al., 2004: 4). In the interpersonal context, social skills can be summarised as involving the ability to express both positive and negative feelings without suffering consequent loss of social reinforcement (Bellack et al., 2004: 4). In essence, for Bellack et al. (2004) social skills involve the ability to observe and analyse subtle cues that define the situation as well as the existence of a collection of appropriate responses. Communication was also considered as a key term when trying to establish a suitable definition of social skills in relation to the school meal. Hargie et al. (1994) identified that a person who is socially skilled is dependent on the extent to which one can communicate with others, in a manner that fulfils one's rights, requirements, satisfactions or obligations to a reasonable degree without damaging the other person's similar rights in a free and open exchange (Hargie et al., 1994: 13). Furthermore, in their review, Hargie et al. (1994) argued that social skills can be acquired through learning and that the control rests with the individual. Hargie (1986) identified six main features of what constitute a social skill, which included 1) goal-directed, 2) interrelated, 3) situational appropriate, 4) identifiable units of behaviour, 5) how behaviours can be learned and 6) control of the individual. Hargie (1986) draws on Bandura's (1977) conceptual framework of social cognitive theory, highlighting the process of social learning.

Another term that emerged when searching for social skills was 'cognition' and this allowed me to extend my review to make links to learning. Bedell and Lennox (1996) emphasised the importance of cognitive elements and identified that social skills included the ability to 1) accurately select relevant and useful information from an interpersonal context, 2) use that information to determine appropriate goal-directed behaviour and 3) perform verbal and nonverbal behaviours that maximise the likelihood of goal attainment and maintenance of good relations with others (Bedell and Lennox, 1996: 9). This suggests that the term social skills represents two sets of abilities: cognitive and behavioural. Behaviour appears to prevail as a term used almost interchangeably with social skills, whether that is positive or negative behaviour. Interaction is also part of the communication process in utilising these social skills.

The 'pedagogic meal'

It is useful to consider how associations to the school meal and learning have been defined and more importantly which term has been used to identify with this phenomenon. It is the notion of the 'pedagogic meal', a concept developed in Sweden in the 1970s which refers to teachers interacting with the pupils when eating school meals, which has been commonly referred to as finding a way to link school food together with learning. There has been a growing interest in the study of food pedagogies (Lalli, 2019; Earl, 2018; Flowers and Swan, 2015; Andersen, 2016; Andersen, 2015; Andersen et al., 2015; Osowski et al., 2013; Pike and Leahy, 2012; Sepp et al., 2006). Sepp et al. (2006) carried out 34 interviews across 12 preschools in Sweden to explore school meals. The staff provided strong views on how food and meals should be integrated into their daily work and pedagogic activities (Sepp et al., 2006). The teachers identified their uncertainties around the 'meal situation' as they lacked knowledge and understanding around food and nutrition. During interviews, participants declared in the past that they did not eat with the children at the dinner table and had difficulties in acting as role models. However, in recent times, staff have been encouraged to socialise with the children at the dinner table even if they are not inclined to do so or been shown how to socialise. The democratic approach to education in Sweden seems to be reflected in their approach to the school meal, which is one of integration and another informal learning opportunity (Mavrovounioti, 2010).

Although most staff in the dining hall had a good understanding of how to practice a pedagogic meal, they remained uncertain of how to present themselves in the meal situation. Sepp et al. (2006) identified how food education occurs early in life for children as they develop preferences for taste, table manners and attitudes towards food. Moreover, it is these attitudes and behaviours that are communicated through the food and meal situation. There is certain behaviour which is seen as acceptable food-related behaviour. Furthermore, they identify how part of this socialisation takes place in preschool tables where behaviours are modelled by staff and pupils.

Overall, it is important to address early childhood and the school meal as this helps to build the foundation of practicing a pedagogic meal. The findings highlighted that staff had a good understanding in encouraging the children to help themselves, as well as acting as adult role models at the table (Sepp et al., 2006: 227); for example, showing the children how to handle cutlery, pass each other food, sit on a chair appropriately and have a conversation. They pointed out how the task of sitting and eating with children and teaching them skills to interact was a task in itself.

Some complexities were identified in studying the meal situation in schooling from previous research carried out in Sweden (Fjellstrom, 2004). These included how social and cultural aspects of children's food habits, including their attitudes behind food choice, have all been important factors in studying from a health perspective (Fjellstrom, 2004: 161). It was also recognised how different dimensions of the meal situation can be observed through looking at time, space and social aspects. It was argued how universal definitions of the 'meal' lacked any idea of 'social dimensions'. By this she meant discussing the meal order, meal patterns and meal situations, which include practices and rituals at the dining table (Fjellstrom, 2004: 161). Fjellstrom (2004) discusses how a 'proper meal' differs amongst Nordic countries, from the structure, daily rhythm and social context of eating. Janhonen et al. (2013) argued that meals that echoed the structural definition of a proper meal were most common when describing meals for the family. Fjellstrom (2004) made an interesting point regarding the relationship between pedagogy and food in a food situation. The social interaction between parents and children in a supermarket has an impact on food purchases and choices in everyday life. This is an example of a pedagogic real-life situation which works as a tool for informal education (Fjellstrom, 2004: 163). It is this notion that is also being carried forward in the school restaurant at Peartree Academy, or at least this was one of the aims of the school.

Observations, interviews and focus group interviews were conducted in three schools in central Sweden to explore how the pedagogic meal is practiced with a focus on teachers' interaction with the children (Osowski et al., 2013). There were three types of teachers identified, who all took different roles, which included 1) the sociable teacher, 2) the educational teacher and 3) the evasive teacher. The sociable teacher created a social occasion during school lunchtime which involved having a high level of interaction with the children. The only difference between the teacher-sociable role and child-sociable role was that the teacher took an interest in the child, giving them attention, meaning that they were able to interact and foster social learning. The educating teacher took the role of providing information during lunchtime, which was a one-way teacher to pupil approach, applying rules and procedures. The evasive teacher took a passive approach, limiting interaction with children and not fully applying rules and procedures. According to the National Food Administration (NFA) as discussed by Filho and Kovaleva (2014) in Sweden, the aims of pedagogic meals are to give children and teachers a chance to interact and speak with each other while eating together and to educate children about food and healthy eating (Osowski et al., 2013: 420). Teachers are seen as role models and the NFA suggests that teachers speak positively about the school meal whilst teaching children about the importance of eating school meals. The Swedish NFA guidelines state that an adult presence brings calm to the school meal environment. Osowski et al. (2013) identified how there was a shortage of literature around the school meal situation and that previous research had merely focused on preschool children.

School interventions can play an influential role in helping pupils to develop social skills as discussed by Durlak and Weissberg (2007). In their study, Durlak and Weissberg (2007) carried out an after-school programme in the United States in an attempt to promote personal and social skills development during mealtime. After-school programmes were defined as interventions that were available for children aged 5–18. Personal and social skills included problem solving; conflict resolution; self-control; leadership; responsible decision making; and enhancement of self-efficacy and self-esteem (Durlak and Weissberg, 2007: 4). Outcomes in three general areas were examined, which included 1) feelings and attitudes, 2) indicators of behavioural adjustment and 3) school performance. It was found that young people who participated in these programmes improved significantly in the three areas. For Durlak and Weissberg (2007) it was possible to identify with these as effective programmes.

The role of school food was considered in connection with pedagogy during school mealtime (Andersen, 2015). Findings from the OPUS school meal project found that staff had a significant impact during mealtimes with regards to their role in both the school kitchen and, whilst it was identified that teachers would be a useful resource during mealtime, the lack of knowledge in terms of how to handle and prepare a basic meal restricted them from taking advantage. This led to a missed opportunity of a pedagogical activity that could have been carried out during mealtime, which could become a formal learning task on food education.

Food education and intervention

Interventions can play a part in enriching the lives of pupils in their participation with school meals as discussed by Burke (2002). Burke (2002) investigated school meals using three focus groups from different post-primary schools, with pupils aged 11–12, based in Northern Ireland. She explored the importance of food and cooking skills by talking to the teachers who taught this subject and canteen staff in the school (Burke, 2002). It was evident that food theory taught in class was not put into practice within the school dining hall. Consequently the low response rate in healthy food choices led to a reduction in healthy food options. Educating children in their food choices was important in engaging with teachers, parents, governors and the wider community. She addressed how the curriculum has the flexibility to be adapted outside the school environment. Schools offer an ideal environment for the development of academic and social skills and also bridge the gap between dietary awareness and food choice. Burke (2002) stressed the importance of the need to attain basic food skills and nutrition in order to make correct food choices in the long-term.

In terms of food intervention, it is worth considering the importance of the roles of those involved. Benn and Carlsson (2014) evaluated the effects of FSM interventions on pupil's learning and on the learning environment in schools in terms of the role of pupils and teachers during the school meal (Benn and Carlsson, 2014). Their research, aimed to explore the learning potential of school meals, was conducted at four schools in Denmark, which generated cases through observations, focus groups and interviews. Overall, they identified that pupils were able to learn through tasting new foods and dishes and argued for the 'common meal', promoting the idea of 'social learning' (Ayers et al., 2007). This was at a different time to the break periods, where pupils brought different food from home. In the lunch period, social learning took place through having and sharing the same meal and experience (Benn and Carlsson, 2014). Furthermore, they also found further learning opportunities through informal learning arenas such as opportunities for pupils and staff to communicate with regards to meal choices.

Health carries cultural, social, moral and linguistic meanings, which is recognised in the research carried out by Karrebaek (2011). An investigation was conducted to highlight the socialisation process of healthy food practices in a Danish multi-ethnic kindergarten classroom based in Copenhagen (Karrebaek, 2011). She adopted ethnographic techniques; observing a class of twenty-five children during a school year by conducting video and audio recordings in class, during breaks and after-school. Karrebaek (2011) focused on the relationship between health, socialisation, language and food practices. Overall, mealtimes were found to expose children to cultural values and act as the platform for them acquiring these values. Children were socialised into a particular society, one which presented them with certain food-related values, including the following: 1) it is important to eat and drink healthy food and drinks and have a specific food-related understanding, 2) milk is healthier than juice and 3) some children have an awareness of and acknowledge the importance of health (Karrebaek, 2011: 16). Overall, it was concluded that careful consideration needs to be given to health interventions in canteens as food practices are built on a number of culture-specific assumptions which cannot be neglected. This emphasises the complexity in trying to prioritise health when considering school meals.

In terms of health, Delormier et al. (2009) suggest that intervention in food consumption should consider eating as a social practice and not an act of behaviour.

To take on the view of a holistic approach then, it is worth considering studies which focus on children's health, nutrition and cognition. Bellisle (2004) identified how diet can affect cognitive ability and behaviour in children and adolescents. She identifies how good regular dietary habits for a child are the best way to ensure they perform well, behave and also gain from benefits in mental health. This particular literature is relevant as Bellisle (2004) emphasises the importance of good nutrition on cognition and behaviour in children, during a crucial period where they acquire factual knowledge, behavioural traits and the social skills that determine their ability to cope with different situations in and outside of the school environment (Bellisle, 2004: 227). For Bellisle (2004) academic achievement and successful integration into a social group depends on numerous factors. These include familial, psychological, emotional, social and nutritional factors. It is these factors which play an important role in the transition young people make into successful adulthood.

Conclusion

The literature review has provided a multi-disciplinary approach to the school meal and highlighted a number of potential issues. In terms of the historical discourse of the school meal, at this present time, school meals seem to be a key priority from a policy perspective. This is reflected in the number of school food programmes being used to find a way to measure the overall mealtime experiences, and the nutritional value of the school meal. I would like to argue that social learning involves using social skills, which are bound by societal rules that impinge upon the participants who inhabit the school restaurant at Peartree Academy.

References

Andersen, S. S. (2015) *School Meals in Children's Social Life: A Study of Contrasting Meal Arrangements*. Ph.D. Thesis, University of Copenhagen.

Andersen, S. S., Holm, L. and Baarts, C. (2015) 'School meal sociality or lunch pack individualism? Using an intervention study to compare the social impacts of school meals and packed lunches from home', *Social Science Information*, pp. 1–23.

Andersen, S. S., Vassard, D., Havn, L. N., Damsgaard, C. T., Biltoft-Jensen, A. and Holm, L. (2016) 'Measuring the impact of classmates on children's liking of school meals', *Food Quality and Preference*, 52, pp. 82–95.

Ayers, S., Baum, A., McManus, C., Newman, S., Wallston, K., Weinman, J. and West, R. eds. (2007) *Cambridge handbook of Psychology, Health and Medicine*. Cambridge: Cambridge University Press.

Bandura, A. (1977) 'Self-efficacy: Toward a unifying theory of behavioural change', *Psychological Review*, 84 (2), pp. 191–215.

Banerjee, R. (2010) 'Social and emotional aspects of learning in schools: Contributions to improving attainment, behaviour and attendance', *National Strategies Tracker School Project*, University of Sussex.

Bauer, K. W., Yang, Y. W. and Austin, S. B. (2004) 'How can we stay healthy when you're throwing all of this in front of us? Findings from focus groups and interventions in middle schools on environmental influences on nutrition and physical activity', *Health Education and Behaviour*, 31 (1), pp. 34–46.

Bedell, J. R. and Lennox, S. S. (1996) *Handbook for Communication and Problem-Solving Skills Training*. London: John Wiley & Sons.

Bellack, A. S., Agresta, J., Gingerich, S. and Mueser, K. T. (2004) *Social Skills Training for Schizophrenia: A Step-by-Step Guide*. New York, NY: Guilford Press.

Bellisle, F. (2004) 'Effects of diet on behaviour and cognition in children', *British Journal of Nutrition*, 92, pp. S227–S232.

Benn, J. and Carlsson, M. (2014) 'Learning through school meals', *Appetite*, 78, pp. 23–31.

Burke, L. (2002) 'Healthy eating in the school environment – A holistic approach', *International Journal of Consumer Studies*, 26 (2), pp. 159–163.

Dalton, T. A. (2004) *The Food and Beveridge Handbook*. Cape Town: Juta Academic.

Delormier, T., Frohlick, K. and Potvin, L. (2009) 'Food and eating as social practice – An approach for understanding eating patterns as social phenomena and implications for public health', *Sociology of Health and Illness*, 31 (2), pp. 215–228.

Department of Health. (2014) 'Think autism: An update to the government adult autism strategy'. Available at: www.gov.uk/government/publications/think-autism-an-update-to-the-government-adult-autism-strategy (accessed: 10 June 2019).

Devi, A., Surender, R. and Rayner, M. (2010) 'Improving the food environment in UK schools: Policy opportunities and challenges', *Journal of Public Health Policy*, 31 (2), pp. 212–226.

DuCharme, R. and Gullotta, T. P. (2012) *Asperger Syndrome: A Guide for Professionals and Families*. London: Springer.

Dudek, M. (2005) *Children's Spaces*. London: Routledge.

Durlak, J. A. and Weissberg, R. P. (2007) 'The impact of after-school programs that promote personal and social skills', Centre for Academic, Social and Emotional Learning (CASEL). Available at: www.lions-quest.org/pdfs/AfterSchool ProgramsStudy2007.pdf (accessed: 2 April 2019).

Earl, L. (2018) *Schools and Food Education in the 21st Century*. London: Routledge.

Filho, W. L. and Kovaleva, M. (2014) *Food Waste and Sustainable Food Waste Management in the Baltic Sea Region*. London: Springer.

Fjellstrom, C. (2004) 'Mealtime and meal patterns from a cultural perspective', *Scandinavian Journal of Nutrition*, 48 (4), pp. 161–164.

Flowers, R. and Swan, E. (2015) *Food Pedagogies: Critical Food Studies*. Surrey: Ashgate Publishing.

Garner, P. (2011) 'Promoting the conditions for positive behaviour, to help every child succeed', National College for School Leadership: Schools and Academies. Available at: http://dera.ioe.ac.uk/12538/1/download%3Fid%D158591%26fil ename%3Dpromoting-the-conditions-positive-behaviour-to-help-every-child-succeed.pdf (accessed: 12 May 2019).

Hargie, O. (1986) *Social Skills Training and Psychiatric Nursing*. London: Croom Helm.

Hargie, O., Dickson, D. and Saunders, C. (1994) *Social Skills in Interpersonal Communication*. London: Routledge.

Hart, C. S. (2016) 'The school food plan and the social context of food in schools', *Cambridge Journal of Education*, 46 (2), pp. 211–231. Available at: http://eprints.whiterose.ac.uk/96751/22/5-18-2016_The%20School.pdf (accessed: 10 January 2019).

Janhonen, K., Benn, J., Fjellstrom, C., Makela, J. and Palojoki, P. (2013) 'Company and meal choices considered by Nordic adolescents', *International Journal of Consumer Studies*, ISSN 1470–6423, pp. 1–9.

Karrebaek, M. S. (2011) 'Understanding the value of milk, juice and water: The interactional construction and use of healthy beverages in a multi-ethnic classroom', *Working papers* in *Urban Language and Literacies*, 83, pp. 1–21.

Lalli, G. (2019) 'School meal time and social learning in England', *Cambridge Journal of Education*. DOI: www.tandfonline.com/eprint/fHGEr7k8ebj55DVBY6YN/full?target=10.1080/0305764X.2019.1630367.

Mavrovounioti, T. (2010) *Democratic education in Sweden*. MA thesis... Lund University, Sweden. Available at: http://lup.lub.lu.se/student-papers/record/1600949/file/1612853.pdf (Accessed: 26 July 2019).

McNeely, C. A., Nonnemaker, J. M. and Blum, R. W. (2002) 'Promoting school connectedness: Evidence from the national longitudinal study of adolescent health', *Journal of School Health*, 72 (4), pp. 138–146.

Nicholson, E. (2005) 'The school building as third teacher', in Dudek, M. (ed.), *Children's Spaces*. London: Architectural Press.

Osowski, C. P., Goranzon, H. and Fjellstrom, C. (2013) 'Teachers' interaction with children in the school meal situation: The example of pedagogic meals in Sweden', *Journal of Nutrition Education and Behaviour*, 45 (5), pp. 420–427.

Pike, J. and Leahy, D. (2012) 'School food and pedagogies of parenting', *Australian Journal of Adult Learning*, 52 (3), pp. 435–459.

Rudd, P., Reed, F., Smith, P. (2008) The effects of the school environment on young people's attitudes towards education and learning, Summary Report, *National Foundation for Educational Research*.

Sepp, H., Abrahamsson, L. and Fjellstrom, C. (2006) 'Pre-school staffs' attitudes toward foods in relation to the pedagogic meal', *International Journal of Consumer Studies*, 30 (2), pp. 224–232.

Weir, C. J. (2008) 'Caterers' experiences and perceptions of implementing the 2006 school meal standards', *Journal of Human Nutrition and Dietetics*, 21 (5), pp. 373–406.

Woolner, P., Thomas, U. and Tiplady, L. (2018) 'Structural change from physical foundations: The role of the environment in enacting school change', *Journal of Educational Change*, 19 (2), pp. 223–242.

3 'You are what you eat'

Learning through school meals

Introduction

School meals could be described as a teaching opportunity, where it is possible for children to learn about acceptable behaviours when eating together, as well as learning about gaining from a positive school meal experience. Such teaching opportunities have been termed as the 'pedagogic meal' in Sweden (Osowski et al., 2013) and this term is applicable to this research as I focus on social learning in the school restaurant. This chapter provides an account of what happens in the restaurant throughout the school day and what social learning occurs. Social learning in this context is described as coming to understand the different discourses that emerge from teaching staff, non-teaching staff and pupils about how to behave appropriately when eating with other people. The school restaurant at Peartree Academy is a place for both meeting and eating and can therefore be regarded as a 'learning space' (Harrop and Turpin, 2013). It provides opportunities for pupils, staff and parents to mingle and socialise (McCulloch and Crook, 2008), and it is also where informal learning can take place (Burke, 2005). However, just as elsewhere in the school, there are rules and regulations in terms of how that social space is used. These rules and regulations in part determine the behaviour of staff and pupils during mealtimes and have an impact upon the social learning that occurs. Data presented in this chapter explores the role of teaching and non-teaching staff and also pupils' views about social learning (Lalli, 2019).

Few studies investigate the social context of the school meal (Lalli, 2019; Hart, 2016; Andersen et al., 2016; Morrison, 1996) and social learning in school dining halls (Osowski, 2013; Jyoti et al., 2005; Ahmed, 2004). Some studies have explored how behaviour and attitudes help shape social skills in schools (Karrebaek, 2011; Durlak and Weissberg, 2007). For example, Karrebaek focused on the relationship between health and school food practices. Social learning is often bound by societal rules and regulations that are created by a community or culture (Sacks and Wolffe, 2006).

There is much confusion regarding the terms social skills and social learning. Social skills are the skills employed when interacting with other people at an interpersonal level (Hargie, 1986: 1). Kelly (1982: 3) adds the dimension of learning by defining social skills as those identifiable, learned behaviours that individuals use during interpersonal situations to obtain or maintain reinforcement from their environment. Knowing how to behave in a variety of situations is part of a social skill. Jyoti et al. (2005) carried out a study on the impact of food insecurity on academic performance and define social learning in a way that allows for educational attainment to be measured. Osowski et al. (2013) indicate that the school meal is a teaching occasion and this definition is used in the context of school dining halls. For Ahmed (2004) school meals were seen as allowing children to develop opportunities for learning in a social way. For me, social learning is about being part of a community and learning about the practical skills of eating and how to behave. The following two themes are discussed in this chapter: 1) what perceptions teaching staff and non-teaching staff have of the restaurant and 2) what social learning occurred in the restaurant.

Perceptions of teaching staff and non-teaching staff

Firstly, the view of teachers is presented and how they understand and interpret social learning. Secondly, data is drawn from non-teaching staff and how they understand and interpret social learning. In some of the data, teaching staff and non-teaching staff refer to it as 'social skills learning'. Therefore, some of the quotations use the phrase social skills rather than social learning. During my observations in the school restaurant, I noticed the divide between the teaching and non-teaching staff in terms of how they were interacting with pupils. I began to question staff understanding of social learning. It also led me to think about discipline as teaching staff were focused on organising the groups of pupils, whereas the non-teaching staff were socialising with pupils. Upon reflecting on the data collected, I noticed how the views of teaching staff and non-teaching staff on what they considered to be social learning differed.

Teaching staff

The views of five teachers, two teaching assistants, pupil guidance leader, the Learning Resource Centre (LRC) leader, an assistant phase leader, senior behavioural leader and the school principal are considered. These staff have a background in teaching. Based on the perceptions of teaching staff, this has been organised under three sub headings: 1) rules: structure and control, 2) building relationships in groups and 3) manners: moral development.

RULES: STRUCTURE AND CONTROL

The teaching staff frequently made links to rules and regulations in relation to the restaurant. Osowski et al. (2013) identified three types of teachers who took part during the school lunch period and identified how it was the educational teacher who led the way for applying rules and procedures whilst interacting with pupils. The comment from assistant phase leader, Charlotte Barry, describes how pupils were subject to time constraints during break and lunch periods. For example:

> Outside of the classroom? Well, it depends what context you mean. There are a lot of extra-curricular activities; there are a lot of clubs, and a lot of things to do, after school particularly. The social times we have here are quite short. We only have about 12 minutes break and 30 minutes lunch. So it's quite a short amount of time then that is literally a quick run around outside, so I think primarily, the movement around the school over lunch time is how pupils get time to interact.
>
> – Assistant Phase Leader

The response from assistant phase leader links to issues of conformity, rules and regulations, which leads to questioning whether the restaurant has been created as a space for this monitoring to take place. She highlighted the structured and controlled nature of the break periods as they were short in duration. Her understanding of social learning was not reflective with the one set out by Dalton (2004) as Charlotte highlights how break time is restricted, i.e. pupil time, which means time for social learning is limited. She describes opportunities for movement around the school as an opportunity for social learning. Social learning involves mutual relations of human beings living in organised communities, and is about being an expert at living, which is essential for getting along with others (Dalton, 2004). In another instance, I observed how the restaurant as a space was being monitored and controlled. For example:

> I have found the school very unique in the sense that the restaurant area is like nothing I have seen before in other schools. There is a level of monitoring and structure going on, but also a lot of integration with the whole school set up. [By this I meant that the school clearly thought through how to utilise space around the school, particularly in the corridors which looked more like learning spaces, not so distinct from a

classroom set up, although I also saw how pupils were being monitored throughout the school day.]

– Field Notes

When I walked around the school, I noticed how in corridors there were spaces for pupils to study outside of the classroom, and these looked as though they were conducive spaces for learning, which is why I thought pupils had opportunities to integrate in the school. However, I also saw how staff controlled pupils throughout the school day and particularly during the lunch periods. This led me to question the purpose of the school restaurant and whether it was a platform for developing opportunities for social learning amongst pupils or about controlling their environment. Based on the evidence in the preceding field note, it can be argued that the restaurant is a space for monitoring and imposing rules.

In relation to being asked about what pupils are learning in the restaurant, the assistant phase leader discussed how to behave and conform to rules. However, she talked about skills which involve using cutlery appropriately and lining up in queues and sitting down to eat properly. For example:

> Well its learning social skills, it's what they should or shouldn't be doing over a lunchtime period, you know how to line up, how to get to the till, how to behave responsibly, how to act with your peers.
>
> – Assistant Phase Leader

Occasionally, I interrupted with follow up questions as it was important to clarify responses, particularly relating to social learning. I asked a follow-up question to the assistant phase leader, who then discussed the importance of instruction in terms of how to help pupils develop social learning in the school restaurant. In her response she talked about lining up, sitting down and learning by example:

> by instruction and example, it's quite different here because we've got the years right from little so they sort of learn it you know the routine of the restaurant from a tiny little dot, lining up and waiting to go and then we've got them sitting down, we sit down, they sit down, and they're quite close to the older year, because we have a staggered lunch time, say for example we have a Year 5, 6 and 8 lunch, with younger pupils as well . . . how they behave, and even the change in the uniforms. The much older years, I think the Year 10 and 11 pupils are independent, I don't think they have younger kids in there, learn by example.
>
> – Assistant Phase Leader

This response from the assistant phase leader demonstrates how the restaurant allows pupils to learn how to behave, although I asked about social learning and she talked about instruction and learning by example. She made the point that pupils learn from one another in the younger year groups. These interactions highlight part of the transitioning periods which take place between year groups. In schools, transition has been identified as an important factor in measuring underachievement (Ekins, 2013). More specifically, she talked about uniforms, routine, waiting, sitting down and a staggered lunch period. According to Charlotte, the restaurant allows pupils to learn from one another.

The teaching assistant described his view on how learning is taking place, through being autonomous, behaving and practicing social skills. For example:

> I think just behaving, showing the way to communicate to people politely, properly you know in a bigger crowd . . . it gives a chance to . . . what's the right way. Show a true picture of yourself.
>
> – Teaching Assistant

The ideas of behaviour and respect appear in the response from the teaching assistent. It is difficult to link these terms to social learning. In this way behaviour is seen as learning. The evidence from the definition by Osowski (2013), Dalton (2004) and Ahmed (2004), coupled with the view of social learning in this chapter, suggests that actually learning how to behave is part of a skill that can be practiced. Furthermore, the teaching assistant highlights a power relationship in the form of behaviour. He is referring to the idea of conformity as pupils are said to be behaving. As discussed by Saldana (2013), schools exercise strategies to ensure pupils are able to conform to the rules that shape the school and for my research, the restaurant is being used as a space for conformity.

The English teacher, Shaun Talbot, also mentioned the term 'behaviour' and particularly emphasised the idea of 'establishing patterns of behaviour', rather than referring to social learning. The question asked was related to social learning. In his view, pupils learn to interact and behave. For example:

> Some of the pupils have difficulty with socialisation and it's trying to get them and to gain established patterns of behaviour, establishing norms for how you should behave in certain situations. So apart from the formal academic side of learning, there are also other patterns of learning going on that feed into their normal learning because we find that some pupils, if they're below a certain level of socialisation, they can't access the learning anyway, because they don't know how to behave in the classroom, they don't know how to speak to adults,

they don't know how to interact with children their own age, so it's all part of it.

– English Teacher

Shaun talked about socialisation as a concept but managed to latch onto key words including 'certain situations', 'how to speak to adults', 'interaction' and 'establishing norms'. In schools there is a growing emphasis on behaviour management, meaning socialisation is perceived in a highly individualised and personalised form (Furedi, 2009: 19). Although, by socialisation, I would like to believe the teacher is talking about social learning. He also made a reference to behaviour and 'patterns of learning', which differ from formal academic learning. He recognised the restaurant as a venue for social learning. The power relationship at play is one of conformity and discipline, which is being exercised in a positive way (Saldana, 2013). This notion of norms and value transmission involves using behaviour management techniques and aligns with his perception; it also offers a potential reason for his association of socialisation with the terms 'norms', 'interaction' and 'how to speak to adults'. For example:

because we find that some pupils, if they're below a certain level of socialisation, they can't access the learning anyway, because they don't know how to behave in the classroom, they don't know how to speak to adults, they don't know how to interact with children their own age.

– English Teacher

It appears that the English teacher confuses socialisation with social learning. He made links to how pupils need to learn certain patterns of behaviour dependent on the situation in which they find themselves. Other research argues that ultimately, teachers are responsible for the pupils' welfare, which includes their physical, emotional and social wellbeing (McCulloch and Crook, 2008: 453). Shaun has a good understanding about social learning and he demonstrates the importance of pupils needing to learn about the eating behaviours which are bound by rules.

BUILDING RELATIONSHIPS IN GROUPS

Findings identified staff associating the restaurant as a place for building relationships. Shaun discussed how bonding between staff and pupils takes place through sociable conversations, before and after lessons. It is useful to consider whether the views of the English teacher on social learning is a more expansive one. For example:

sociable conversations, nothing to do with work, so that's an opportunity to see people and that building up relationships really helps, so preschool and after school that helps bonding between staff and pupils.

– English Teacher

The perspective of Shaun Talbot is a significant one as he offers an in-depth view of social learning. He makes a reference to 'sociable conversations' and it is useful to consider one view of how a sociable conversation is interpreted. For Simmel (1950), in a sociable conversation the topic is not important; it is about developing a charm and attraction for talk.

A Year 7 teacher describes how pupils are not just learning about developing relationships, but also about looking after their own wellbeing. It is useful to draw on the idea of a whole-school approach which was identified as a way to address pupil wellbeing and the approach was to promote pupil connectedness (Banerjee, 2010). For example:

> there are social skills clubs, even relationships with staff are promoted, in terms of greeting each other in corridors, and discussing things, not just about the pupil as learning but everything about the pupil making sure that their wellbeing is at the forefront which has an impact on their education.
>
> – Year 7 Nurture Teacher

She identified the importance of developing relationships in order to learn social skills. I also observed how pupils were developing relationships when conversing in groups. For example:

> There are a number of Year 7 pupils chatting away and acknowledging the group sitting opposite them which I've identified as a group of Year 9 of pupils [although different Year groups are in at the same time, they are still divided by Year group and seated accordingly]. It is interesting to note how the Year 9 pupils have also come in closer together and chatting away and the Year 7 pupils are now also talking about football. Is this a common interest area? Or is this an opportunity to develop social learning through conversation?
>
> – Field Notes

The topic of conversation I observed plays a huge role in bringing pupils together, as the topic of football is a shared interest as recorded in the preceding field note. Based on Ahmed's (2004) view, school meals are seen as allowing children to develop in a social way. It could be argued that the school restaurant is the space in which they are able to have a dialogue. For instance, if this conversation were to take place in the playground, it would be interesting to consider whether the conversation would last as long or whether it be overtaken by more physical activity. Although pupils do not necessarily have to sit together, it is the common ground upon which they become connected that they are able to interact and converse. It could be argued that the restaurant encouraged pupils to share these types of discussions which lead to social learning.

Roy Piston, an academic maths coach, talked about relationships and provided another view of social learning. His perception is based on a comparative view of other schools, although he did not provide any specific examples of how other schools operate. He described the relationship between pupils and staff and suggested that it is not so formal, and how they consume lunch together. It is said that pupils and staff who dine together are able to develop their social skills and behaviour (DfE, 2007). Roy draws on his ideas in the following example:

> I find that relationships between students and staff there's a divide usually, you know the students talk to the staff and the staff talk to the students. It's much more easy going here, I will certainly stop to say hello to any students I come across. Sometimes we sit together and have lunch, teachers and students will sit downstairs and have lunch together.
> – Academic Maths Coach

Staff are modelling how to build relationships in the preceding example. However, Roy was not able to elaborate with specific examples and therefore did not offer any depth in his view of social learning.

One teaching assistant shared a similar view to the academic maths coach but described how pupils are able to use the restaurant to sit together in groups. School dining is said to bring groups of pupils together and create an opportunity for them to socialise (Stevens et al., 2008). She described how pupils use the restaurant to learn how to be part of a group and makes reference to social learning. For example:

> They're learning how to be part of groups, they tend to sit with their friends, and the social skills are there.
> – Teaching Assistant

For the teaching assistant, pupils are learning to work with one another in a group setting during lunch, and she states how pupils tend to sit with friends, but does point out how social learning is taking place in this situation.

The Year 7 nurture teacher highlighted how conversations that take place between staff and pupils lead to relationship building where these groups are able to mix together in the restaurant. For example:

> conversations between staff and pupils, and the fact that staff and pupils mix together, sometimes you'll get groups of staff sitting together, but

quite often you'll see staff integrating together, with the children . . . modelling the social skills of the general how's the day gone and what have you been up to, what you doing tonight? I think that helps them quite a lot, compared to literally across the road where staff don't interact with pupils.

– Year 7 Nurture Teacher

She was referring to the primary school across the road, although she did not explicitly say which school she was referring to in the response. Furthermore, she did not mention visiting the primary school. There seems to be a tendency to compare to the school across the road. She seems to have only a partial understanding of what social learning entails. She makes associations to modelling behaviour when talking about social skills or social learning. It is interesting to consider why she perceives social learning in this way. Her implicit if partial understanding is confirmed by the work of Hendy and Raudenbush (2000), who found associations between teacher modelling as being one of the most effective methods of encouraging food acceptance in children.

MANNERS: MORAL DEVELOPMENT

Another common theme from the findings was able table manners and moral development, which is about recognising the importance of teaching pupils about how to behave at mealtime whilst modelling good eating behaviours as discussed by Birch (1980) and Eliassen (2011). In the following example, one teacher points out how saying 'please' when in the restaurant is a form of social learning. For example:

> *Pupils are also correcting one another around the table, e.g. 'Pass me the water . . .', 'Say please!' Pupils are also analysing their puddings [just comparing with one another, saying hmmm, mine looks nicer than yours]. A Year 6 pupil has been asked by a female member of staff to return to his side of the restaurant as this section belongs to Year 8 pupils.*

STAFF (TEACHER) – 'See . . . that is a social skill!' (*Looking in my direction.*)

– Field Notes

In the preceding example, pupils are in discussions which link to manners in a number of ways. By commenting on each other's puddings, they are exercising positive eating behaviours. More specifically, this links to the work by Rahim et al. (2012) and Eliassen (2011), who make links between school food and manners in terms of how pupils are able to exercise good manners during mealtime. This teacher's knowledge of my research and how I am concerned with social learning also begins to surface as she looks towards me for reassurance. Therefore, this piece of data indicates that manners, including please and thank you, are understood by this teacher as a form of social learning.

In the following response from the science teacher, another view of social learning is presented. In this case, he talked about the displays on the wall: respect and forgiveness as forms of moral development. It is useful to consider changes to the school dining environment which were discussed by Rudd et al. (2008), who found that pupils responded positively to changes to the dining environment which included changes to the aesthetics. The study by Rudd et al. (2008) considered the views of pupils before and after the renovations. There is something to be said about the wall displays in the school restaurant at Peartree Academy, and links between social learning and wall displays were made by the science teacher in support of this notion. For example:

> they can learn social skills . . . which are also displayed on the walls. You've got posters, when they're eating, most time, they're facing the dinner ladies or dinner men, whoever is serving, they've got stuff to read around to do with respecting you know, forgiveness and they can read around that area. [Unsure whether there will be any further evidence which will allow me to evidence pupils actually reading these posters.]
>
> – Science Teacher

He assumes that the pupils actually read the posters and did not simply see it as wallpaper. During my time at the school, I did not notice these posters ever changing and I did not hear staff encouraging pupils to read them; or at least I do not have any record of them doing so. While the intentions of the school may have been to encourage moral and social development through the poster displays, my evidence does not support this as an effective approach. For me, this active modelling of behaviour was a more appropriate position in terms of social learning.

The principal discussed a number of factors in relation to social learning in the school restaurant, but here, she presents a similar response to that of the deputy principal, who also talked about manners and moral development synonymously with social learning. The principal described the responses of visitors to the school and the manners pupils display when visitors are present. For example:

> well we all know that skills around even using a knife and fork, manners, developing healthy eating skills, developing social skills . . . . I have lots of visitors to this academy, and I can tell you now that

visitors that come from outstanding schools tell me, that our Year 10 and 11 in particular are exceptional in their manners, they can sit there for a good 20 to 25 minutes and they are impeccably well behaved because they've learnt how to use social time in that way. I think whatever we're thrown in teaching people manners, in teaching people everything, in that restaurant at lunch time, it serves up every opportunity.

– Principal

The principal talked about manners and how visitors perceive the restaurant. The point that she described 'exceptional manners' as being able to sit in the restaurant for twenty to twenty-five minutes, whilst using a knife and fork properly, suggests that firstly pupils are learning how to behave, but it also highlights a power relationship. The argument she presented does have value but it is also due in part to disciplinary pupil behaviour. She referred to skills pupils acquire, making reference to pupils learning about manners and suggesting that pupils are being controlled in their environment. By discipline, I am referring to the mechanism of power which regulates the behaviour of pupils (O'Farrell, 2005).

Pupil guidance leader Marilyn Huston described how pupils are learning about social skills, manners and also how to eat. She discussed how she encourages pupils to help them develop autonomy; developing positive eating behaviour as discussed by Eliassen (2011). The pupil guidance leader provided her perception of how pupils are using the restaurant for social learning in this way. For example:

they're learning social skills . . . also manners, and eating skills cutting up food and just general table manners . . . I'm in charge of a team of midday supervisors, and I always encourage them to show how to set a good example to the children. I always encourage them to help them with the food but to encourage the children to do it for themselves, help them learn.

– Pupil Guidance Leader

She highlighted how pupils are encouraged by midday supervisors to set good examples, particularly in terms of food choices and table manners. This is also discussed by Eliassen (2011), who links eating behaviour to role modelling.

The head of English described how pupils are learning about the social skills of eating and using cutlery appropriately. For example:

I think they learn the social skills of eating together, sitting at a table, using the cutlery, that kind of thing, eating proper food, eating a rounded meal, which many children in lots of walks of life but particularly some of our pupils who haven't had a meal at home, they don't necessarily sit at the table, and eat as a family.

– Head of English

She also mentioned how some pupils do not have the opportunity to eat together at home, but did not elaborate. She assumed that pupils are not eating together at home which links to the work of Bergh (2014), who identifies with the importance of trying to create an environment which is similar to the home.

The deputy principal identified how pupils are able to learn how to use the right cutlery. He pointed to mealtime experiences for some of the pupils, who he claimed did not have access to a 'proper meal'. Janhonen et al. (2013) provided one view of a proper meal, which was considered as structural and most common when describing meals for the family. H identified how pupils were using lollipop sticks to pick up the chicken breast instead of using a fork. For example:

> early on I saw some kids, 12, 13 year olds, with a chicken breast with a fork and lollipop stick because they're not used to using a knife and fork . . . so all those are good things because we get a lot that, they don't have tables at home, get them to sit around a table and actually have proper home cooked meal.

– Deputy Principal Finance and Resources

According to the deputy principal, pupils did not always have access to a dining table at home and were not familiar with how to use a knife and fork. Whilst it is difficult to find evidence to support his claim, he does identify with the school restaurant as a space for interacting, although cutlery skills in children have been identified as a problem area (Piercy, 2008).

The senior behavioural leader was also in agreement with the deputy principal and highlights how some of the older pupils often struggle with using cutlery. It is said that with the shift towards a fast food culture, children have not been able to develop cutlery skills (Piercy, 2008). The senior behaviour leader draws on his experience of how this is affecting pupils. For example:

> We've got some pupils in Year 10 who aren't sure how to use a knife and fork.

– Senior Behaviour Leader

Lucy Tyler, the assistant principal, also described how pupils struggled to use cutlery and pointed out how the school spent a considerable amount of time teaching pupils about table manners prior to her appointment at the school. For example:

> before I came they spent a lot of time actually teaching children to have those table manners really and a lot of the children weren't able to use knives and forks properly.
>
> – Assistant Principal

Therefore, there is evidence to suggest that pupils are learning the very basic skill of how to use cutlery in the school restaurant. This emphasises the importance of the role of food in education (Kaklamanou et al., 2012).

The assistant principal discussed how pupils are learning about politeness and presented with opportunities to make healthy food choices. Andersen et al. (2016) highlight how children's food choices are affected by the way in which the food environment has been developed. Lucy described how the restaurant allows for the opportunity of sharing food together. For example:

> I think through the restaurant and through the opportunity to share food. I know at the beginning it was a real focus for the academy, so there was a really big focus . . . encouraging them to be polite really and social to each other at the table and then we have lunch time supervisors as well who again just make sure children are making good choices with their food, finishing their lunch and just sorting out any issues that they're having over the lunch table.
>
> – Assistant Principal

This point illustrates how the restaurant encourages pupils to be polite and social although this is not described any further. She identifies how lunchtime supervisors ensure pupils are making good food choices whilst encouraging them to be polite.

The senior behaviour leader discusses how pupils are able to learn about what is acceptable behaviour and social cues, particularly how manners are being acquired. For example:

> Well like I said, I think, you touched on it earlier, the key social cues that they may not already have, so it's that wider life skill learning you know, the social cues, what's acceptable in a restaurant, what's acceptable around a dinner table, how you conduct yourself around other people when you're eating you know . . . they're key social skills that everybody needs. And also then you are teaching them about different

foods and what, it's an opportunity to you know say why you should eat this food and model these behaviours.

– Senior Behaviour Leader

He also draws on the importance of teaching pupils about how to behave during mealtime and modelling eating behaviours (Birch, 1980; Eliassen, 2011). He makes links with what to eat as well as how to eat as important factors all contributing to a healthy school experience. He identifies the responsibility of both staff and pupils, who are in a position of modelling social behaviour by eating with each other, minimising behavioural difficulties whilst developing opportunities to interact in a social way.

The pupil guidance leader highlighted the importance of politeness, manners, walking sensibly and following the rules. Whilst this is also about positive behaviour, it is the disciplinary actions that are being implemented. For example:

just talking to their peers as well as the adults in a polite manner, not just shouting across, talking in small groups, being polite to adults and listening, doing as they're asked, walking sensibly.

– Pupil Guidance Leader

The LRC leader discussed social learning but linked to the difference between 'right' and 'wrong'. It is known that most schools now operate with the assumption that teachers are able to take on the role of the parent in order to teach them about values (Aspin 2007: 48). For example:

it's a lot down to social skills in the restaurant, what sort of table etiquette, what's right what's wrong.

– LRC Leader

There are two lines of argument in what was said by the LRC leader. Firstly, that the restaurant is a place where social learning takes place in the form of learning about 'table etiquette', but then secondly, she makes the link to right and wrong. The evidence suggests that she is interested in the moral development of pupils and refers to this as a form of social learning. It is about 'expertness' and the ability to get along with other pupils in the school restaurant. Dalton (2004) describes how this expertness emerges through practised ability and she highlights this when talking about achieving independence in pupils.

Overall, it is evident that staff perceive social learning in different ways. The correlations gleaned from this part highlight the perceptions of the teaching staff. When discussing social learning, teaching staff seem to be referring to three common themes, which include rules, conformity and manners. Therefore, rather than talking about pedagogy, teaching staff seem to be predominantly concerned with monitoring behaviour and rules. The next part attempts to provide an account of how non-teaching staff perceive social learning.

Non-teaching staff

In this part, the views of an exams officer, a catering manager, a midday supervisor and the LRC leader are presented. I discovered that both the exams officer and LRC leader did not hold formal teaching qualifications. The reason why responses from these individuals have been included is due to their closeness to the school restaurant in terms of their daily roles. Based on the perceptions of non-teaching staff, this has been organised into three discussions: 1) manners: dining etiquette, 2) control: a safe environment and 3) communicating together.

MANNERS: DINING ETIQUETTE

Mealtimes are said to be particularly important for children so that they can be taught to eat correctly whilst using good table manners and making use of cutlery properly. This is said to help children to develop confidence in any type of environment with a food setting (O'Sullivan, 2015). Lomax (1999) identifies the school meal experience as a special communal space and time of day for sharing, which includes placing importance on table manners, conversation, togetherness and community.

The catering manager described her view of how pupils learn to adapt to the rules of etiquette in the school restaurant. For example:

> Even from 3 years old, we decided, that plates would get broken, but then their learning on a plate from the word go, you know and it was a good thing. And the little ones, well I thought they wouldn't be able to carry food . . . but yes they can, they'd line up just like everyone else to go to a separate hall, place.
>
> – Catering Manager

The following example from my field notes highlights an interaction between a non-teaching staff member and pupil, which focuses on instruction via the use of a knife and fork:

> *One member of staff, who is sitting next to a pupil, is explaining how to use a knife and fork.*

STAFF – 'The fork is held in the left hand with the knife in the right!"'
PUPIL – 'Oh, are you sure?'
STAFF – 'Look at him over there' [*staff member points to another staff member*].

– Field Notes

In this example the staff member is providing instructions for the pupil in the school restaurant; this is evidence of an observation where I was able to capture one type of social interaction between pupil and staff. Rahim et al. (2012) carried out an investigation on school meals and found that parents were particularly keen for pupils to develop social skills, table manners and dining etiquette. The preceding example from the field note involves a more competent individual, showing how to perform a specific task to a pupil.

The catering manager pointed out how some of the pupils lack an awareness of how to handle cutlery being in the restaurant. For example:

> The social skills, I think from our point of view basically, in the dining room, where from the word go, we wanted it to make it homely, we wanted to put flowers on the tables, we wanted to make it bright and open to encourage people to come into it, to do meetings, which was the main focus. Basically, a lot of children were not aware of how to use a knife and fork.

– Catering Manager

She has a particular view about what is meant by social learning and social skills. She suggests it is about creating a community space that is homely and welcoming and will encourage children to learn simple tasks. Therefore, based on the catering manager's view, the restaurant allows for social learning to take place through the teaching of practical skills.

CONTROL: A SAFE ENVIRONMENT

The LRC leader made links to school meals and the need for a safe and controlled environment. Lomax argues how time spent in the school dining area is a time for interaction, sociability and self-discipline (Lomax, 1999). This creates a sense of belonging where pupils feel safe, yet controlled so that behaviour can be monitored and pupils are able to have a structured day. In addition to this, Ahmed (2004) argues that social skills or social learning involves allowing pupils to develop in a social way, with the school meal experience being instrumental in potentially allowing for this to happen. For example:

> breakfast, it's one of the important meals of the day, often our pupils come in to school without having had anything to eat and then expecting

them to work and concentrate, they were expecting too much of them. So, providing that breakfast and for them to have something to eat, and also, you know a lot of them have been kicked out the house early on, and if their parents work, it is somewhere safe and warm for them to go as well, which also gives them structure

– LRC Leader

Rochelle, the LRC leader, talked about the restaurant as an environment in which pupils can be safe and secure and nurtured. She also made a link between breakfast and learning. She suggested that pupils cannot be expected to work and concentrate if they are hungry. Greaves et al. (2016) points out how school food is one of the ways in which learning is influenced. This view is further supported by the work of Stevens et al. (2008), who found a link between school breakfast and how it helped pupils achieve better academic results. In addition to this, Benn and Carlsson (2014) evaluated the effects of FSM intervention and saw how this had an effect on pupils learning outcomes.

The Year 7 teacher suggested that the restaurant is a safe environment for pupils to socialise with peers whilst eating breakfast. It is said that children have an important role in healthy school initiatives, such as the food provided in the school canteen, which should be a clean and safe physical environment (Taylor et al., 2012). For Sheila, the Year 7 teacher, pupils are able to come to the restaurant, but this is separated from the 'learning environment', which I infer to be the classroom. Therefore, Sheila does not see the restaurant as a formal learning environment but does see it as a safe environment for developing friendships. For example:

where they can come to a safe environment before learning [classroom], chat to their friends while they're having breakfast . . . I do think that's a really big bonus.

– Year 7 Teacher

She stated that the time spent before learning is a 'big bonus' for pupils and she related specifically to the period before the morning lesson. By using the term safe environment, she is suggesting that the restaurant is a controlled space in which pupils are able to develop social learning opportunities.

COMMUNICATING TOGETHER

The exams officer identified the relationship between learning, social skills and behaviour. She talked about the importance of developing autonomy in pupils through communication, although I had to probe to get her to explain what she meant by 'develop social skills'. For example:

> Well again it's the social skills, they're learning how to develop those and . . .
>
>   well learning to work and communicate with peers that are older than them or adults how to behave independently, how to get their own meals.
>
> – Exams Officer

Her response suggests that she has a better grasp of social learning compared to the teaching staff. To an extent, it can be argued that she does, as teaching staff neglect any mention of the idea of independence and personal development. For Flutter and Rudduck (2004: 133), it is important to recognise and respect pupils' need for autonomy in learning and instil them with a sense of independence. There seems to be a contrast between how teaching staff understand social learning compared with the non-teaching staff. The midday supervisor also discusses communication and how it plays a role in shaping opportunities for learning. For example:

> yeah it's good, they're learning about communicating together in the restaurant by sitting with each other, whether they are 10 or 14 years old.
>
> – Midday Supervisor and Parent

The non-teaching staff made reference to communicating together, dining etiquette, manners and control in the school restaurant. Whilst teaching staff also made reference to these areas, they tended to focus predominantly on rules and regulations. On balance, it is the non-teaching staff who appear to have a better understanding of social learning and are ultimately the group that are seen to be ironically teaching pupils specific skills and encouraging social learning. Teaching staff do not appear to be as knowledgeable as the non-teaching staff with regards to social learning; there seem to be some discrepancies in their understanding of social learning. Furthermore, when I asked teaching staff about social learning, they referred to rules, working together and manners, linking closely to behaviour. Whilst Dalton (2004) would argue that rules are important in getting along with others in society, the restaurant as a space for developing social skills could be a good pedagogical learning opportunity. However, some of it is being missed because the staff do not understand what sort of social learning could be encouraged.

What social learning occurred in the school restaurant?

Learning spaces are defined as non-discipline specific spaces visited by both staff and pupils for self-directed learning activities, which are commonly within and outside of library spaces (Harrop and Turpin, 2013).

The idea of learning spaces has been studied in connection with learning outcomes (Kuuskorpi and Gonzalez, 2011; Nair and Gehling, 2008; Durlak and Welssberg, 2007; Burke, 2002). According to the OECD (2006), educational space is physical and supports multiple and diverse teaching and learning programmes and pedagogies; a place which encourages social participation. The learning space in question relates to the school restaurant. Burke (2002) discussed the role of the school meal as a place for informal learning, which she termed as the 'edible landscape of school'; a place and time of day associated with the consumption of food and drink (Burke, 2002: 573). This is the notion that informal learning opportunities are being presented in areas of the school other than the classroom.

The relationship between the pupil and the meal is embedded in the everyday life of the school. According to Benn and Carlsson (2014), it can be an occasion for relaxation and informal learning through the meal experience; it can also be a more organised and structured aspect of the school day. This discussion is made up of two segments, 1) staff and 2) pupils.

Staff

This part presents the views of the following staff who discuss what pupils are learning in the school restaurant: five teachers, two pupil guidance leaders, a teaching assistant, a senior behavioural leader, a catering manager, an assistant principal and the deputy principal. This has been split into five segments: 1) socialising and interaction, 2) a space for group teaching, 3) food choices: 'knowledge', 4) practical learning via coaching and 5) patrolling, controlling and setting the standards.

SOCIALISING AND INTERACTION

The data analysis highlighted the following terms: 'socialising' and 'interacting'. By this I mean the restaurant is seen as a potential space for allowing pupils to interact together with various year groups. As discussed by Filho and Kovaleva (2014), the idea of a pedagogic meal aims to give children and teachers the chance to interact. Michaela Perry, the English teacher, presented an account of how the school restaurant is used as a space to allow pupils to mix together, particularly during break and lunch time. For example:

> I think during break time and lunch time obviously they're encouraged to mix in kind of communal areas rather than find little cubby holes to hide away and then outside which is monitored by staff. I think that improves interaction.
>
> – English Teacher

Michaela also made the point of pupils being encouraged to interact in communal areas rather than find little cubby holes to hide away; she then stated how pupils are monitored outside by staff. From this, it seems the staff are encouraging pupils to use the restaurant but there is a power relationship as all pupils do not necessarily want to be using the space, but the reason for not wanting to use the space is not clear. The school meal is known to have different types of stigma attached to it, one of which includes take up of FSMs as pupils do not always like the association (Storey and Chamberlain, 2001). The more apparent power relationship identified in the response by this teacher is where she referred to the monitoring of pupils taking place by staff.

The Year 7 teacher (all subjects) described the school meal as a social occasion which allows for sharing to take place. He discussed how opportunities are presented for interacting and sharing meal experiences. It is said that in contrast to the classroom environment, the school dining hall offers a period of non-instructional time that pupils view as a much more flexible social occasion (McCullough and Hardin, 2013: 176). In the following example, the Year 7 teacher, Simon Woolley, described how pupils are learning to interact with each other:

> well they're learning how to interact with each other, that's the main thing, they're learning how to sit down in a social occasion and share an event like a meal, it's the only time they sit down and have a meal and actually do sit down and share a meal with others because for whatever reason, that might not be happening at home. So I think the social aspect of sitting down together in a group and also because we don't have a separate staff dining room or you know an area like that. You'll often find members of staff siting with the children as well, and it's a good way to build and form relationships sometimes, if you find it tricky inside the classroom.
>
> – Year 7 Teacher (All Subjects)

In this example, Simon discussed the importance of the social aspect of the school meal as discussed by McCullough and Hardin (2013). He identified how the restaurant is being used as a social learning area and said that the main idea behind the development of the restaurant is for pupils to sit down in a social occasion and share a meal. Benn and Carlsson (2014: 25) discussed how the meal was the most communal human act and that the arrangement of a communal meal is part of a cultural and social lesson. For Delormier et al. (2009), eating should be considered as a social practice as opposed to an act of behaviour. Furthermore, Simon also emphasised how there is no separate dining hall for staff, who also sit with the children, and pointed out how the restaurant is a good way of building relationships. Pupils are learning about interacting and socialising with both peers and staff in the restaurant according to the views of both Michaela Perry and Simon Woolley.

A SPACE FOR GROUP TEACHING

Another theme that emerged from the data collection was the emphasis on the school restaurant as a space for group teaching. As discussed by Wills et al. (2015), the dining hall is increasingly becoming known as both a space for teaching and dining. English teacher Shaun Talbot described how the restaurant is a useful platform for teaching pupils, referring to it as a 'learning area' and a place where informal learning can take place. For example:

> the restaurant is used as a learning area as well, so I teach there . . . for a number of lessons every week and also see a lot of teachers doing similar things. The learning area, sometimes EAL other times with English pupils doing work. I see a lot of teachers doing that and its formal academic teaching but it's also being together and being sociable and eating together, they are learning other things, they are learning social skills and I think that's one of the main drivers to creating the restaurant to allow them to be socialised.
>
> – English Teacher

In his response, the English teacher identified how he is using the restaurant as a formal and informal way to teach pupils and placing an emphasis on being sociable. He teaches in the restaurant for a number of lessons but pupils are allowed to be sociable and eat together. He highlighted the importance of the restaurant as a formal and informal learning space, by suggesting that pupils are learning other things whilst eating together and highlights how the restaurant functions as the space to allow pupils to be socialised. This is supported by Osowski (2013), who states how the school meal is another teaching occasion. Furthermore, Dimbleby and Vincent (2013) support the notion that the school dining hall is an informal space in which the whole school is able to come together.

Alan, the deputy principal for behaviour and safety, discussed how the restaurant creates an almost 'naturalistic learning' opportunity for pupils to interact with one another in a group. For example:

> they get obviously what I call a natural opportunity with their friends, they're in the restaurant with their friends, in any kind of new group situation, all kind of together and have that interaction . . . there are some success activities designed to go on to encourage sometimes there run by people who have got skills in that area, they'll patrol that environment.
>
> – Deputy Principal Behaviour and Safety

He suggested that the restaurant allows for pupils to develop what he terms a 'natural' opportunity to develop friendships. The restaurant has the potential to create certain learning opportunities for social learning to take place which is also about getting to know new peer groups. The second point Alan identified is how the restaurant is managed by staff who have a particular skill set and those are the staff members who patrol the restaurant.

The teaching assistant discussed how pupils are presented with opportunities to socialise in the restaurant and learn to become part of groups. For example:

> we do loads of after school clubs, there's loads of ways to get together, more social side of skills is the restaurant, it's our main port of call, so everyone mingles in there, and they're allowed their phones out and the school with the WIFI, they're learning how to be part of groups, they tend to sit with their friends, and the social skills are there.
>
> – Phase 3 Teaching Assistant

She highlighted how internet access is available for pupils and that this provides them with opportunities to learn in groups. She mentioned the importance of learning to be part of a group, and that this is encouraged by being allowed to use their mobile phones. This is essentially another rule as pupils are only allowed to use their mobile phones at particular times. Learning about being part of a group is about knowing when to be a contributor and when to take the leading role, whilst listening to others' opinions (Arnoff and Wilson, 2014).

Therefore, the deputy principal behaviour and safety, English teacher and teaching assistant recognise the school restaurant as a space for teaching which takes place through formal group work in terms of teaching and also establishing new groups of friends in an informal way, which links back to social learning.

The teaching staff highlighted the importance of food choices and this was a theme which emerged from the data analysis, about building knowledge and good food choices in the restaurant, whether that involves the food itself or how to behave when eating. For Roberts and Marvin (2011), children's knowledge towards healthy eating has an impact on their food choices and attitudes to food. Lucy, the assistant principal, emphasised how food is used as an avenue to teach pupils how to make good food choices. For example:

> I think things have improved in the children, social skills have improved, there's not as much of a need but it's still there as an opportunity . . . a lot of the children, it's surprising how hungry they are when they come in and . . . It's great to have it there because often if a child is having any issues one of the first things we'll say is do you need something to eat?
> – Assistant Principal

For the assistant principal, the restaurant allows for social learning opportunities to take place; she also stated how staff prioritise asking pupils for food. Lucy makes an assumption that parents are also making poor food choices, which lead to children taking on these eating behaviours. This notion is supported by Eliassen (2011), who identified how families were seen as children's first important role models of eating behaviour.

The assistant principal described her experiences of how the restaurant promotes certain opportunities, including the encouragement from staff to be polite and socialise with one another around the dining table. She also described her view on how the restaurant brings about opportunities in making food choices and building knowledge. Food choices are also discussed by Hamilton-Ekeke and Thomas (2007), who point to education as playing a key role to changing the attitudes and behaviour of children in order for them to make good food choices. Lucy also distinguished between 'structured time' and 'unstructured time', which is the move from the classroom to either the restaurant, playground or other parts of the school. She introduced the role of the lunchtime supervisors, who are there to intervene with potential issues amongst pupils. For example:

> They're learning about healthy food choices . . . they're learning to be sociable with each other, there is often a lot of disagreements in the restaurant, but the minute you go out of that structured time from the classroom and you move into the unstructured time, as teachers we have to spend a lot of adult resources, getting the children to walk down . . . sensibly and . . . come back up sensibly . . . when the lunch

time supervisors come in and the teachers leave, so there's a lot of them learning to work with different adults, learning to be respectful, to an adult that's not their teacher.

– Assistant Principal

She described how the lunchtime supervisors take over in place of the teachers and pointed out how opportunities are created for pupils to interact with various members of staff. She differentiated between learning to work with an adult who is not their teacher and learning to be respectful. She described how pupils are able to interact with adults in the school restaurant. She talked about making healthy food choices, structure, walking sensibly which is also discussed by Hamilton-Ekeke and Thomas (2007). This was observed in my field notes. For example:

Pupils seem to be well mannered, as they walk through the restaurant, asking to be excused as they pass by the pupils [this is something I noticed quite regularly, pupils seemed to have respect for the space outside the classroom].

– Field Notes

In the example from my field notes, pupils are behaving appropriately in the restaurant whilst exercising good manners, so pupils seem to have respect for the space outside of the classroom. Therefore, in this view the school restaurant had been created in a way that meant pupils would respect the space. Rudd et al. (2008) also considered the school dining space and highlighted how pupils reacted positively to changes made to the dining hall when questioned before and after the renovation. However, when pupils were standing in queues, I observed something quite different. For example:

Pupils begin taking out frustrations, using foul language to vent on route to the buffet, specifically, Year 9 pupils. The pupil finishes her comments with 'Excuse my language!' [So the pupil recognising that using bad language is breaking the rules of the restaurant.]

– Field Notes

The assistant principal spoke about getting pupils to walk sensibly, which is about appropriate behaviour. The preceding field note highlights one example of how bad manners and bad choices are still in evidence.

The inclusion leader made reference to how pupils are trained how to use a knife and fork, a similar response to the one set out by the deputy principal and senior behavioural leader. This is about attitudes and behaviours towards eating habits which is discussed by Roberts and Marvin (2011).

The inclusion leader also suggested that certain foods are not prohibited in the school restaurant. For example:

> They're told how to use a knife and fork, all those that bring in their lunch box, it has to be a healthy lunch box, no sweets are allowed in there and no fizzy pop and if that's seen they'll just confiscate it straight away.
>
> – Phase 1 Inclusion Leader

She draws on the rules and regulations set out by the school on what pupils are allowed to bring in their school lunch boxes. This is about teaching them and their parents to make good food choices. While it could be argued that pupils are encouraged to eat healthy food, it might also be argued that pupils are just doing what they are told to avoid getting into trouble. They are not necessarily learning why sweets and fizzy pop are not healthy food choices (Pike and Leahy, 2012).

PRACTICAL LEARNING VIA COACHING

In a response from Lana, the catering manager, she draws on the pedagogical approaches in terms of teaching methodologies, taking a reflective approach in trialling strategies to integrate food, learning and the space within which the activity is carried out. She introduced the context through storytelling and it was this narrative that enabled pupils to engage and taste new foods whilst making use of the restaurant to learn. Furthermore, Lana also talked about how food production is about developing the pupil's knowledge about food in a practical way through coaching. For example:

> I tell you what we do . . . I do every first term where I have the children come in and they all learn about our job. And last week was my job . . . they come in, I used to talk in the classroom, but I found they wanted to see what was going on, so what they do now is they come in, one in the morning, one class, this is the 5 year olds, it was really successful, they see the food production, I introduce them to everybody, show the children what they're making and then in the afternoon we do cakes, and then we show them that and we go to the table with little questions and then we have a sample of what we made, then we made flapjack, cookies and they love doing that. I also do another one where they're learning about the story goldilocks and the three bears, so I make them porridge and then we talk about the story of goldilocks and how it works. You know the sort of things to introduce different things.
>
> – Catering Manager

The catering manager highlighted how pupils are able to learn to cook and participate in a question and answer activity around the table in the school restaurant. Adopting a pedagogical approach (Osowski et al., 2013) during mealtimes was not established from the beginning, according to the catering manager, who disclosed how the transition was difficult in trying to integrate everyone to sit and eat together, including staff and pupils. For example:

> they didn't do that at the beginning, at the beginning they sat with the older children at the top of the dining hall, it was a case of them [the younger pupils] being pushed out the way, so let's keep them separate. The teaching assistants and teachers come in and have meals with them to make sure they eat their food, so it's like a community, they all sit together.
>
> – Catering Manager

It would appear that the teacher involvement was structured to ensure teachers mingled with all the pupils. The catering manager identified the relationship between the staff and pupils and how practical learning opportunities are presented. Overall, it could be argued that pupils are encouraged to engage in practical learning opportunities in the school restaurant.

PATROLLING, CONTROLLING AND SETTING STANDARDS

Much of the staff rhetoric is about the restaurant as a community social space but it became evident that it is a space that is both highly controlled and 'patrolled' by teachers. The deputy principal highlights how the restaurant is patrolled by staff. For example:

> they'll patrol that environment, sometimes; particular groups are set up to look at the needs of those children.
>
> – Deputy Principal Behaviour and Safety

The pupil guidance leader, Chloe, discussed how the restaurant is used as a space for pupils to set standards for one another. This is about the restaurant as a space for encouraging good behaviour and staff are able to point out the good behaviour of another pupil (Bentham, 2012). Chloe highlighted how staff are expected to perform and also set standards. This is about teacher modelling, which has been identified as one of the most effective methods of encouraging food acceptance in pupils (Hendy and Raudenbush, 2000). For example:

> Well, they're learning from others aren't they, and you know looking up to their elders, because a lot of the phase 2 children have their lunch at 12.15.

. . . Year 7, 8 and Year 9 are in the restaurant at the same time so they kind of get to look up to them and get to see what expectations are within the restaurant and things like that. We've also got staff members on duty as well, so we know what's expected of them and standards are set really.

– Pupil Guidance Leader

She discusses expectations, standards and duty as three key terms in relation to being able to learn from others in the restaurant. Overall, the narrative in this part supports the interpretations of Kuuskorpi and Gonzalez (2011), who studied the physical makeup of learning environments. They discussed how informal learning was taking place away from the traditional classroom. They argued that advances in communication and information technology have contributed in shaping the way teaching operates in schools today.

It also supports the work of Nair and Gehling (2008), who discussed the experiences of young people within the confines of a classroom, arguing that they do not reflect the diversity of settings and relationships young people must learn to negotiate in order to succeed in the workplace (Nair and Gehling, 2008: 11). They argued the divide between formal and informal learning activities is a divide which largely relies on the design of the school building. They also drew on school cafeterias in their account and described them as a place for refuelling, with little time or space to socialise or collaborate on projects. This means that potentially, the school restaurant at Peartree Academy could be a space for bridging that gap between formal and informal learning activities, so that it is more than just a place for 'refuelling'.

This part was about the perspective of staff, who highlighted their perception of social learning within the restaurant. It presented examples of how pupils are learning about making food choices, how to behave and modelling different behaviours. There were some correlations between the perceived idea of learning and what pupils actually learnt in the restaurant. Overall, power relationships were discussed and the subheadings reflect the idea of control in the restaurant, whether that is a positive view (through instruction and learning about manners) or negative view (controlling pupils and enforcing them to use the restaurant).

Pupils

This part presents the views of pupils and what they consider is being learnt in the school restaurant. Pupils referred to two common themes when discussing their views of the school restaurant, which included 1) organisation of the restaurant and 2) rules and expectations. There are both positive and negative perceptions of the pupils presented in this final part of the chapter. The pupils include the following: Year 11 pupil × 2 (15 years of age), Year 10 pupil (13 years of age), Year 9 pupil × 2 (12 and 13 years of age), Year 7 pupil (12 years of age), Year 5 pupil (10 years of age), Year 6 pupil × 2 (11 years of age) and Year 4 pupil × 3 (aged 8 and 9). The reason why responses from these particular pupils are selected is due to their interest in the layout of the school restaurant.

ORGANISATION OF THE RESTAURANT

Pupils at Peartree Academy made consistent reference to the organisation of the school restaurant. Rudd et al. (2008) carried out an investigation to explore the impact of pupil perceptions on changes to the school dining environment. The work of Rudd et al. (2008) is useful as it highlighted positive responses from pupils following a revamp of the school dining hall in terms of the way it had been renovated and managed. It was the day-to-day running of the restaurant and how the tables and chairs are set out in particular that were a part of pupil comments in the restaurant. One Year 11 pupil described the layout of the restaurant and was not sure opportunities for learning were being presented. For example:

> I don't like the way it's laid out, there's only a few circle tables, then you got long tables, you've got massive groups . . . we don't understand why only adults are sat there on circular tables.
>
> – Year 11 Pupil

In this example, she highlighted her perception of the hierarchical layout of the tables, as the circular tables are placed in the middle of the restaurant and she mentioned how adults predominantly occupy these tables. There is a power relationship in the restaurant and she pointed out how only adults were allowed to be seated on the circular tables. This is about staff presence and a strategy for ensuring staff are present in the school restaurant. It has been known for quite some time that the school lunch period is the biggest behaviour-related problem in schools and that discipline is paramount (Lomax, 1991: 55). This pupil offered a rich account of the restaurant compared to other pupils who were not able to provide as much detail. This conflicts with the views presented by staff of the restaurant which all seem to be positive, describing it as a community and social space.

Four other pupils also made reference to the organisation of the school restaurant, sharing similar views, paying particular attention to the lack of space available. McNeely et al. (2002) examined the association between school connectedness and the school environment, which highlighted the importance of taking a whole-school approach. The reason for this approach was to ensure that schools would keep the health and wellbeing of a pupil in mind. All four pupils made the comment that the restaurant would function better as a space if it were bigger in size. For example:

I wish the restaurant could be bigger.

— Year 6 Pupil

it's like too crowded cos the tables are too much together, hard to get out.

— Year 6 Pupil

I don't like it because its crammed and packed, so it's like going in at once, sometimes the card machines don't work so have to take on cards [by this the pupil is referring to the vending machines which are operated by a smart card].

— Year 9 Pupil

Things that I don't like about it, it's like too crowded sometimes, they take the water before you use it, some tables they take the cups and leave the jugs out.

— Year 11 Pupil

The views of the pupils in the preceding examples seem to be conflicting with the views of staff. Pupils seem to be discussing negative aspects of the restaurant whilst staff did not highlight anything which involved hygiene, space or crowdedness. Burke and Grosvenor (2003) recognised the importance of the significant role of eating for young people but also recognised the negative connotations which is said to involve an unsettling time and space for pupils. The following five pupils discussed the day-to-day organisation of the restaurant, more specifically identifying issues of hygiene in terms of both the restaurant and food being served. For example:

I don't like sometimes food is getting out of date, a bit mouldy.

— Year 4 Pupil

I don't like when floors are not clean.

— Year 4 Pupil

when they clean the tables they use the same cloth over again, the dirty food that's gone on to the cloth they use the same one to clean it.

— Year 5 Pupil

I don't like the hygiene; it's disgusting like cups have got all dirt in them and everything.

– Year 7 Pupil

the glasses are really dirty, have the water, the waters always warm so not cold and I don't like that and that's it.

– Year 10 Pupil

There were no comments from staff in terms of the hygienic state of the school restaurant but it is important to note that pupils were asked about what they disliked about the restaurant. Children are more likely to tell the truth and this is one of the reasons why I was able to gather responses in which negative perceptions on the restaurant were presented. Whilst I did not record any specific examples of issues surrounding hygiene (see the following field note) in the school restaurant, I did notice sticky glasses with water marks left on them, but this could be down to the usage of the restaurant and difficulty in maintaining the cleanliness of the space. For example:

I am yet to see a dirty table or a pupil sitting at a table that hasn't been wiped.

– Field Notes

I noticed how queues were not as short as discussed by staff and this was evidenced in a field note. For example:

Year 9 pupils begin to stroll in as noise levels seem to decrease with a calmer atmosphere. Staff members have lined up with pupils as queues become larger [so why are staff saying queues are relatively short when generally, pupils seem to think the opposite?].

– Field Notes

Whilst there seems to be positive comments from staff, pupils seem to view the restaurant in a completely different light. For example, pupils are saying the queues are quite long although there seems to be a general consensus amongst staff that queues are very short.

RULES AND EXPECTATIONS

It was evident from the data analysis that pupils were interested in discussing rules and expectations in the restaurant. It is useful to consider what type of rules and expectations are imposed upon pupils. As discussed by Saldana (2013), the school exercises strategic power relations as a means and as an end to teaching conformity and in this case the school restaurant is being used as the platform for discipline. This is about conformity and one Year 10 pupil and one Year 11 pupil outlined how expectations are imposed upon them in the restaurant in the following example:

> The environment's alright, we're told to make sure we sit properly and behave for the younger lot . . . the Year 9s just go in as we have dinner.
> – Year 11 Pupil

> the line for the dinner, sometimes . . . I don't like when you don't have much time . . . have 10 minutes to have your lunch [the pupil was referring to how the queue was quite long for the lunch period and the punishment was that lunch had to be consumed in a short period due to a lack of uniform on the day].
> – Year 10 Pupil

I did not observe a period where pupils were being punished in this way. Pupils appear to be being disciplined for not wearing a school uniform. In terms of learning about rules, this pupil described her experiences in the restaurant with the staff. For example:

> the cleaners that wipe tables, ask if you're ok, actual dinner ladies, evil, don't like any of them, she dropped a plate once, so I was like have you dropped your favourite plate and she was like don't talk to me, and wasn't happy. Don't like the dinner ladies here, they should be getting the sack not the behaviour support team, can cook our own dinners.
> – Year 11 Pupil

For Burke and Grosvenor (2003), the negative connotations attached to the school meals usually consisted of a perception of school canteens as noisy places, often dull with robotic dinner assistants, including midday supervisors. The restaurant at Peartree Academy has been designed in a way which allows those who step into the restaurant to be monitored at any time of day. In order to unpack what is being learnt, it is important to consider how pupils perceive the restaurant. A Year 4 pupil highlighted how the restaurant is not offering an appropriate salary for her mother for the duties she carries out on a daily basis, as a midday supervisor in the restaurant. For example:

> she's not like paying them the right amount, my mom does the after-school clubs and she don't pay her right. She's getting like a low amount of money.
> – Year 4 Pupil

There is a level of expectation here and this was evident from other responses in which pupils tend to highlight areas of concern with regards to the restaurant.

A more positive view of the restaurant is presented by other pupils. Pupils from Years 4, 6 and 9 described how opportunities are presented to work as part of a team as well as the restaurant being a space for spending time with friends. For example:

> Pretty good 'cos they do nice food and always help each other, no one at the desk who will not help.
> We've got the older children to learn what to do and how to be a good pupil.
>
> – Year 4 Pupil

> Like you can hang out there with your friends and talk until the bell rings.
>
> – Year 6 Pupil

> It's a useful space for everyone really . . . activities, working together, meeting up and obviously eating as well.
>
> – Year 9 Pupil

These views of the pupils coincide with the views of the majority of staff. One interpretation that can be made is that pupils are learning about how to work as part of a team through observation. They are learning about interacting and maintaining relationships with peers and friends. In addition, they recognise the importance of helping one another and modelling behaviour of other pupils in learning how to be a good pupil.

The pupils described how they are learning to work as part of a team and interact with one another and also about making good food choices. Learning is said to be based on a result of collaborative intention, and one pedagogical method is to use the school restaurant as a potential space for doing so (Osowski, 2013). I would like to argue that the school dining area is a platform for both imposing rules upon pupils whilst also a potential space where social learning can take place. Other responses from pupils were directed towards the hygienic element and organisational side of the restaurant which involved a discussion on the condition of the food and lack of space in the restaurant. There is evidence to suggest that the restaurant can foster opportunities for social learning, but there is a complex array of variables involved in making this judgement.

Conclusion

It is clear that there are positive aspects in which the school restaurant is helping to foster opportunities for social learning, although there is a pressure which is holding this social learning from taking place at times. When referring to social learning, teaching staff make reference to four common areas. These included 1) rules and regulations, 2) how to behave, 3) relationship building in groups and 4) manners: moral development. Non-teaching staff made reference to three common areas, which included 1) community, 2) manners: dining etiquette and 3) communicating together. Pupils were more interested in discussing issues of organisation, rules and hygiene. There is evidence to suggest that learning opportunities are presented in the school restaurant and this was evident in the second part of this chapter where views of both staff and pupils were discussed. For pupils, it was evident in some places that social learning was taking place, although they were more inclined to highlight issues of hygiene and how there was a lack of space available.

Overall, this chapter presents an argument to suggest that non-teaching staff have a better understanding of social learning and make little reference to rules and regulations compared to the teaching staff. So, it is important to consider what could be done to help tackle this missed opportunity for social learning. One argument is that schools should prioritise the dining space so that it is conducive for social learning. There are still differences in perception between teaching and non-teaching staff as well as the pupils. There was consensus amongst pupils on particular issues which highlighted the positives of the restaurant but conversely, some pupils were in agreement with the negative aspects which were preventing the restaurant from working as a platform for fostering social learning opportunities.

References

Ahmed, A. U. (2004) *Impact of Feeding Children in School: Evidence from Bangladesh*. Washington, DC: International Food Policy Research Institute.

Andersen, S. S., Vassard, D., Havn, L. N., Damsgaard, C. T., Biltoft-Jensen, A. and Holm, L. (2016) 'Measuring the impact of classmates on children's liking of school meals', *Food Quality and Preference*, 52, pp. 82–95.

Arnoff, J. and Wilson, J. P. (2014) *Personality in the social process*. London: Routledge.

Aspin, D. N. and Chapman, J. D. (2007) *Values education and lifelong learning: Principles, policies, programmes*. London: Springer.

Banerjee, R. (2010) Social and emotional aspects of learning in schools: Contributions to improving attainment, behaviour and attendance, *National Strategies Tracker School Project*, University of Sussex.

Benn, J. and Carlsson, M. (2014) 'Learning through school meals' *Appetite*, 78, pp. 23–31.

Bentham, S. (2012) *A Teaching Assistant's Guide to Child Development and Psychology in the Classroom*, 2nd ed. London: Routledge.

Bergh, A. (2014) *Sweden and the revival of the capitalist welfare state*. Cheltenham: Edward Elgar Publishing.

Birch, L. L. (1980) Effects of peer models' food choices and eating behaviours on pre-schoolers' food preferences, *Child Development*, 51, pp. 489–496.

Burke, C. and Grosvenor, I. (2003) *The School I'd Like: Children's and Young People's Reflections on an Education for the 21st Century*. London: Routledge Falmer.

Burke, L. (2002) 'Healthy eating in the school environment – A holistic approach', *International Journal of Consumer Studies*, 26 (2), pp. 159–163.

Dalton, T. A. (2004) *The food and Beveridge handbook*. Cape Town: Juta Academic.

Delormier, T., Frohlick, K. and Potvin, L. (2009) Food and eating as social practice – an approach for understanding eating patterns as social phenomena and implications for public health, *Sociology of Health and Illness*, 31 (2), pp. 215–228.

Department for Education. (2007) Learning environments for pupil referral units, London: Stationary Office.

Dimbleby, H. and Vincent, J. (2013) The School Food Plan, *Department for Education*. Available at: http://www.schoolfoodplan.com/wp-content/uploads/2013/07/School-Food-Plan-2013.pdf (Accessed: 10 August 2019).

Durlak, J. A. and Weissberg, R. P. (2007) The impact of after-school programs that promote personal and social skills, *Centre for Academic, Social and Emotional Learning (CASEL)*. Available at: http://www.lions-quest.org/pdfs/AfterSchool-ProgramsStudy2007.pdf (Accessed: 2 August 2019).

Ekins, A. (2013) *Understanding and tackling underachievement*. London: Optimus Education.

Eliassen, R. K. (2011) 'The impact of teachers and families on young children's eating behaviours'. Available at: www.naeyc.org/files/naeyc/Eliassen_0.pdf (accessed: 10 February 2019).

Filho, W. L. and Kovaleva, M. (2014) *Food waste and sustainable food waste management in the Baltic sea region*. London: Springer.

Flutter, J. and Rudduck, J. (2004) *Consulting pupils: What's in it for Schools?* London: Psychology Press.

Furedi, F. (2009) *Socialisation as behaviour management and the ascendancy of expert authority*. Amsterdam: Amsterdam University Press.

Greaves, E., Crawford, C., Edwards, A., Farquharson, C., Trevelyan, G., Wallace, E. and White, C. (2016) Magic Breakfast: Evaluation report and executive summary, Education Endowment Foundation: Institute of Fiscal Studies. Available at: https://educationendowmentfoundation.org.uk/pdf/generate/?u=https://education endowmentfoundation.org.uk/pdf/content/?id=963&e=963&t=Breakfast%20 clubs%20found%20to%20boost%20primary%20pupils%E2%80%99%20 reading%20writing%20and%20maths%20results&s=&mode=embed (Accessed: 11 February 2019).

Hamilton-Ekeke, J. T. and Thomas, M. (2007) 'Primary children's choice of food and their knowledge of balanced diet and healthy eating', *British Food Journal*, 109, pp. 457–468.

Hargie, O. (1986) *Social Skills Training and Psychiatric Nursing*. London: Croom Helm.

Harrop, D. and Turpin, B. (2013) 'A study exploring learners' informal learning space behaviours, attitudes and preferences', *New Review of Academic Librarianship*, 19 (1), pp. 58–77.

Hart, C. S. (2016) 'The school food plan and the social context of food in schools', *Cambridge Journal of Education*, 46 (2), pp. 211–231. Available at: http://eprints.whiterose.ac.uk/96751/22/5-18-2016_The%20School.pdf (accessed: 10 January 2019).

Hendy, H. M. and Raudenbush, B. (2000) 'Effectiveness of teacher modelling to encourage food acceptance in preschool children', *Appetite*, 34 (1), pp. 61–76.

Janhonen, K., Benn J., Fjellstrom, C., Makela, J. and Palojoki, P. (2013) Company and meal choices considered by Nordic adolescents, *International Journal of Consumer Studies*, ISSN 1470 – 6423, pp. 1–9.

Jyoti, D. F., Frongillo, E. A. and Jones, S. J. (2005) 'Food insecurity affects school children's academic performance, weight gain and social skills', *American Society for Nutrition*, pp. 2831–2839.

Kaklamanou, D., Pearce J. and Nelson, M. (2012) 'Food and academies: a qualitative study', Research Report, *School Food Trust, Department for Education*. Available at: https://www.education.gov.uk/publications/eOrderingDownload/Food%20and%20Academies%20-%20a%20qualitative%20study.pdf (Accessed: 17 February 2019).

Karrebaek, M. S. (2011) 'Understanding the value of milk, juice and water: The interactional construction and use of healthy beverages in a multi-ethnic classroom', *Working papers in Urban Language and Literacies*, 83, pp. 1–21.

Kelly, J. A. (1982) *Social Skills Training: A Practical Guide for Interventions*. New York, NY: Springer.

Kuuskorpi, M. and Gonzalez, N. C. (2011) 'The future of the physical learning environment: School facilities that support the user', *Centre for Effective Learning Environments CELE Exchange, 2011/11*. Available at: www.oecd.org/edu/innovationeducation/centreforeffectivelearningenvironmentscele/49167890.pdf (accessed: 3 February 2019).

Lalli, G. (2019) 'School mealtime and social learning in England', *Cambridge Journal of Education*. DOI: www.tandfonline.com/eprint/fHGEr7k8ebj55DVBY6YN/full?target=10.1080/0305764X.2019.1630367.

Lomax, P. (1991) *Managing Better School and Colleges: The Action Research Way (BERA Dialogues)*. Bristol: Multilingual Matters Ltd.

Lomax, P. (1999) 'Working together for educative community through research', *British Educational Research Journal*, 25 (1), pp. 5–21.

McCulloch, G. and Crook, D. (eds.) (2008) *The Routledge International Encyclopedia of Education*. London: Routledge.

McCullough, M. B. and Hardin, J. A. (2013) *Reconstructing obesity: The meaning of measures and the measure of meanings*, Oxford: Berghahn Books.

McNeely, C. A., Nonnemaker, J. M. and Blum, R. W. (2002) 'Promoting school connectedness: Evidence from the national longitudinal study of adolescent health', *Journal of School Health*, 72 (4), pp. 138–146.

Morrison, M. (1996) 'Sharing food at home and school: Perspective on commensality', *Sociological Review*, 44 (4), pp. 648–674.

Nair, P. and Gehling, A. (2008) 'Democratic school architecture', *Voices in Urban Education*, Annenberg Institute for School Reform, 34. Available at: http://vue.annenberginstitute.org/sites/default/files/issuePDF/VUE19.pdf (accessed: 3 February 2019).

OECD (2006) CELE Organising Framework on Evaluating Quality in Educational Spaces, Organisation for Economic Co-operation and Development. Available at: http://www.oecd.org/education/innovation-education/evaluatingquality ineducationalfacilities.htm (Accessed: 4 August 2019).

O'Farrell, C. (2005) *Michel Foucault*. London: Sage Publications.

Osowski, C. P., Goranzon, H. and Fjellstrom, C. (2013) 'Teachers' interaction with children in the school meal situation: The example of pedagogic meals in Sweden', *Journal of Nutrition Education and Behaviour*, 45 (5), pp. 420–427.

O'Sullivan, J. (2015) *Successful leadership in the early years*. London: Featherstone Education.

Piercy, N. (2008) *Market-led strategic change*. London: Routledge.

Pike, J. and Leahy, D. (2012) 'School food and pedagogies of parenting', *Australian Journal of Adult Learning*, 52 (3), pp. 435–459.

Rahim, N., Kotecha, M., Callan, M., White, C. and Tanner, E. (2012) 'Implementing the Free School Meals Pilot', RR228, *Department for Education*. Available at: https://www.education.gov.uk/publications/eOrderingDownload/DFE-RR228.pdf (Accessed: 10 March 2019).

Roberts, K. and Marvin, K. (2011) *Knowledge and Attitudes Towards Healthy Eating and Physical Activity: What the Data Tell Us*. Oxford: National Obesity Observatory.

Rudd, P., Reed, F. and Smith, P. (2008) 'The effects of the school environment on young people's attitudes towards education and learning', *Summary Report*, National Foundation for Educational Research.

Sacks, S. and Wolffe, K. E. (2006) *Teaching social skills to students with visual impairments: From theory to practice*. United States: AFB Press.

Saldana, J. (2013) 'Power and conformity in today's schools', *International Journal of Humanities and Social Science*, 3 (1), pp. 228–232.

Simmel, G. (1950) *The Sociology of Georg Simmel*. New York: Simon and Schuster.

Stevens, L., Oldfield, N., Wood, L. and Nelson, M. (2008) 'The impact of primary school breakfast clubs in deprived areas of London', School Food Trust, Eat Better Do Better.

Storey, P. and Chamberlain, R. (2001) 'Improving the take up of free school meals' Research Brief 270, London: DfE.

Taylor, N., Quinn, F., Littledyke, M. and Coll, R. K. (2012) *Health education in context: An international perspective on health education in schools and local communities*. London: Springer.

Wills, W., Draper, A. and Gustafsson, U. (2015) *Food and public health: Contemporary issues and future directions*. London: Routledge.

4 'Come dine with me'
Surveillance mechanism or community forum

Introduction

This chapter questions whether the school restaurant is a community forum or another mechanism for surveillance. The overall argument presented highlights what actually goes on in the restaurant as opposed to its original intended purpose, which is to foster opportunity for social learning. From reflecting on the life and culture of Peartree Academy, clearly whilst the revamping of the dining facilities helps to make the school meal a more attractive occasion, it is important to take advantage of this forum which can help towards promoting social learning. It is a space in which training takes place in shaping future citizens. Therefore, two narratives are highlighted in this chapter, running in parallel: firstly that the restaurant allows pupils to converse and interact, acting as a community forum, and secondly that there are certain pressures working against this which leads to surveillance by staff. In support of the restaurant as a community forum, it is useful to look closely at eating behaviours. It is useful to consider the notion of eating behaviours and to recognise how the term behaviour is bound by societal rules and regulations. This led me to develop a discussion on the management of the school restaurant alongside the modelling of behaviours (Lalli, 2019; Pike and Leahy, 2012; Danaher et al., 2002).

Eating behaviours

In this part I am using the term 'eating behaviours' to refer to how pupils behave and what they choose to eat. When discussing eating behaviours, food pedagogies are particularly important because of the social, cultural and symbolic meanings of food and potential 'good lives' that can be produced (Flowers and Swan, 2015). Flowers and Swan (2015) were referring to the influence of food and how it can help young people to 'do good' and 'be good'. By this, they are referring to the influence of eating behaviours on being a 'good member' of society. Therefore, food is seen by society as a means through which we can improve our individual and collective lives (Flowers and Swan, 2015: 19). There are a number of studies which have investigated eating behaviours (Osowski et al., 2013; Eliassen, 2011; Richards and Smith, 2007; Sepp et al., 2006) and it is useful to highlight what was found and how they interpreted the term eating behaviours. Osowski et al. (2013) highlighted the notion of the pedagogic meal (Sepp et al., 2006) and identified three different roles staff were modelling during children's mealtime experiences, which include 1) sociable teacher, 2) educational teacher and 3) evasive teacher. Although it can be argued that children's eating is a private and family affair, they are taught what, when and how to eat and become subject to educational regulation at school (Osowski et al., 2013). The work of Osowski et al. (2013) is particularly useful for my study as I also noticed how staff were modelling different behaviours in the school restaurant at Peartree Academy.

Eliassen (2011) argued how children were dependent on their teachers and families to support their wellbeing and promote positive development, including eating habits. This study is useful in that it considers the influence of both staff and parents in terms of the impact on eating habits. Richards and Smith (2007) investigated the factors that influenced food choices for children and identified how it was staff in schools and family members who played a significant role in shaping their attitudes and values towards food by acting as role models. This is useful as it lends itself to the narrative of eating behaviours with a link to role models. Sepp et al. (2006) carried out a study to identify staff members' attitudes to the role of food and meals as part of daily activities in a school. They identified how attitudes towards interaction played a key role in children's eating experiences and this relates closely to my study as I also make an attempt to frame social learning in this way.

In terms of 'eating behaviours', the definition that is being adopted for my study is the one put forward by Eliassen (2011), who states that like the family, teachers are able to model eating behaviours in schools which involve social interactions with children at mealtimes. So, for Eliassen (2011), eating behaviours are defined as behaviours set by educators and parents and it is the behaviours they model and social interactions during mealtimes that are of key importance in ensuring children are able to gain from this. Eliassen (2011) also discussed how positive role modelling around children would help to ensure a positive attitude is shaped towards food. Whilst Eliassen (2011) is more concerned with food choices, I would like to use the part of the definition which places an emphasis on modelling of eating behaviours and social interactions to help explain my own research.

Surveillance

This idea links to the second narrative running throughout this chapter which is based on the notion that the restaurant is used as a space for monitoring pupil behaviours. Firstly, it is useful to look at what is meant by surveillance (Danaher et al., 2002; Foucault, 1980) before drawing on some influential studies (Saldana, 2013; Punch et al., 2013; Pike and Leahy, 2012; Pike, 2010) which have helped inform the discussion in this chapter.

There is evidence to suggest that the discourse of surveillance has always been part of the structures in society as Foucault wrote about these ideas in 1980 and 22 years later, Danaher et al. (2002) are still discussing them. Surveillance and self-regulation techniques have become a fundamental part of life in western societies, particularly in spaces like shopping centres where security cameras have been installed (Danaher et al., 2002). Surveillance has become an attribute of modern society in institutions like schools. The definition carried forward in this chapter is the one presented by Foucault (1980) who states there is no need for arms and violence, just a 'gaze', where each individual (pupil or staff) exercise surveillance over and against themselves (Foucault, 1980: 155). The notion of visibility has been used in part one of this chapter to highlight how this is happening.

I would suggest that pupils are being monitored in the restaurant and that there is a sense of surveillance shaping the day-to-day running of the restaurant. I am concerned with the fragilities between establishing whether pupils are able to take up social learning opportunities or whether surveillance hinders this. It is said schools have a role to ensure learning takes place in a safe environment and also that schools must provide excitement, challenge and discipline (Calvert and Henderson, 1998: 15). This narrative aims to argue how pupils are being monitored in the form of a 'learner' vs. 'trainer relationship', i.e. pupil vs. staff (Calvert and Henderson, 1998). The learner is the pupil and the trainer becomes the staff member who is monitoring pupils in the school restaurant.

The restaurant was originally designed to help foster opportunities for social learning. However, in addition to the many activities described earlier in this thesis, the restaurant as a space seems to be used by staff to watch over and monitor pupil behaviours. For example:

> A female member of staff is constantly monitoring every move of the younger pupils, using hand gestures to calm them down [it seems that there is a level of patrolling going on in the restaurant].
>
> – Field Notes

Having provided a definition of surveillance, it is useful to look at power relations which shape mealtime. In order to make sense of what is going on in the restaurant, it has been useful to use some key studies (Saldana, 2013; Punch et al., 2013; Pike and Leahy, 2012; Pike, 2010) which make links to notions of surveillance and how this shapes the school meal.

Saldana (2013) argues that a school exercises strategic power relations as a means and as an end to teaching conformity and that some pupils learn to become agents in its services, whilst others learn to oppose it (Saldana, 2013: 228). My work also draws on issues of how strategies are being implemented in order to monitor the day-to-day running of the school restaurant. Punch et al. (2013) adopted a Foucauldian lens to investigate the cultures of school dining halls and the ways in which social relationships are constructed and reconstructed. This is useful as the restaurant is also a place where fragile relationships are being negotiated. Pike and Leahy (2012) studied discourses of food pedagogies and reveal how pedagogical techniques of surveillance are deployed by schools and by this they are referring to lunchbox surveillance. I am interested in exploring the notion of surveillance and Pike and Leahy (2012) identify how the school meal is bound by surveillance. This links back to my research, as one teacher said that fizzy drinks and sweets were taken out of lunchboxes at Peartree Academy. Pike (2010) argues that school dining spaces are neglected areas of inquiry and that they should no longer be regarded as a 'passive container' for human activity. It is evident that research on school mealtime and surveillance is a recent area of inquiry in which investigations have begun to highlight the complexity in interpreting the school mealtime.

The rest of this chapter has been divided into two parts: 1) managing the school restaurant and 2) modelling eating behaviours in the restaurant. The first part explores how the restaurant is managed, drawing on discussions of the staggered lunch break, pupil participation, control and structure. The second section highlights a discussion on modelling eating behaviours in the restaurant. This includes a discussion on the idea of the school building as the 'third teacher', which refers to how the school environment can help shape learning interactions (Nicholson, 2005). It is said the role modelling of healthy food choices can have a massive impact on children and can involve a range of activities including cooking healthy food, preparing better snacks, and having dinner with the family whilst promoting a positive experience, further promoting positive associations with healthy foods (Garvis and Pendergast, 2017: 142). Furthermore, there is much to be said about the role modelling of eating behaviours and in some instances, this involves making the right food choices (Birch, 1980). It is argued that the school restaurant can act as a platform for training and inducting people.

Overall, this chapter investigates the fragilities between whether this was achieved or whether the restaurant became another platform for surveillance. The first part introduces a discussion on the management of the school restaurant.

Managing the school restaurant

Only some research has focused on how behaviour is managed in school dining halls (Pike, 2010). This part of the chapter introduces a discussion on how the school restaurant is managed, which supports the notion of the restaurant as a surveillance mechanism. Power can be organised at minimal cost to the school as it can exist in the school structures (Foucault, 1980). By this, I am referring to how staff are using the school restaurant to control pupil behaviours. Sub-sections to follow include 1) the staggered lunch break system, 2) pupil participation and 3) structure and control. The key focus of the first part is to highlight how the staggered lunch break system works. There is a problem in terms of how the shorter lunch break hinders opportunities for social learning to take place. The second part introduces a discussion on pupil participation in the restaurant and the negotiations which take place in terms of how some pupils conform with the short lunch break whilst others choose to resist (Saldana, 2013; Danaher et al., 2002). The final part explores how structure and control is exercised in the form of staff presence in the school restaurant.

THE STAGGERED LUNCH BREAK SYSTEM

One issue that impinges upon the school meal is the length of lunchtime (School Food Trust, 2007). This part highlights the logistics of the school lunch which includes an insight into the timings by year group. It also highlights the rationale behind the introduction of the staggered break system and describes a normal lunch hour in the UK. The School Food Trust (2007) developed a document to support schools in finding ways of organising a suitable lunch break system and a staggered lunch break is one form that is mentioned.

At Peartree Academy, due to the number of pupils, seating them all at the same time is not possible and therefore the school decided to stagger the lunch break. The school adopted this system in order to allow all year groups to have their lunches and also to minimise behavioural disruption, so this one approach was adopted for a dual reason. The school lunch period begins at 11.30 am running through until 2.15 pm and this is becoming normal practice in large UK schools (Lightfoot, 2007). Data from my field notes explains how the lunch break is arranged. For example:

Having bumped into one of the administrative staff, I was told how the lunch break was staggered over two hours and fifteen minutes. There is also some overlap [by this the administrative staff member was referring to the fact that sometimes things could over run].

Staggered break time:

11.30 am–2:15 pm (School lunch period)

– Field Notes

The younger pupils, i.e. year groups 1–3, ate in a separate dining hall, whereas the rest of the school ate in the school restaurant. In terms of the timings of the typical school day, there is no statutory requirement for the length of the break times and it is up to the governing body to decide. Lightfoot (2007) reported that it was uncommon for schools to have one-hour lunch breaks due to behavioural difficulties that might arise but recognises the importance of pupils needing time to sit and eat with peers whilst discussing issues of the day. At Peartree Academy whilst the lunch break is spread over two hours and fifteen minutes, each year group has 40 minutes to get to the restaurant, queue up for lunch, eat and is then required to leave the restaurant in order to allow the other year groups to enter. The rest of the lunch break is spent either in other parts of the school or in the playground. Shorter lunch breaks have been identified as being a worrying trend in schools and this has led to the questioning of the negativity placed upon pupils as a result of this (Zandian et al., 2012). Traditionally, the school lunch period lasted for 60 minutes, more commonly referred to as the lunch hour (Devi et al., 2010). At Peartree Academy there are three main lunchtime sittings which include eating time as well as playtime. This has been illustrated in a chart (see Table 4.1).

Based on the chart, 40 minutes for a lunch break seems relatively short, bearing in mind that pupils are required to fit in quite a lot during this short

Table 4.1 Illustrating lunch break timings

Year group	Time	Duration
Primary Years 4, 5 and 6.	11.30 am–2.10 pm	40 minutes
Restaurant break: catering staff clear restaurant ready for the next year group.	12.10 pm–12.35 pm	25 minutes
Secondary Years 7 and 8.	12.35 pm–1.15 pm	40 minutes
Restaurant break: catering staff clear restaurant ready for the next year group.	1.15 pm–1.35 pm	20 minutes
Secondary Years 9, 10 and 11.	1.35 pm–2.15 pm	40 minutes

period. It was interesting to learn about why the school lunch break had been reduced. One of the English teachers discussed the rationale behind the introduction of the staggered lunch break and pointed out how it was reduced to minimise behavioural issues:

> We have a shorter lunch period . . . I don't know if you know the background of it, because we're experiencing difficulties with pupils who have too much time on their hands. That was one of the reasons for reducing time available. So, it's all staggered because we have to get the whole school through but previously there used to be a longer period. We're finding that pupils, once they'd eaten, once they'd had that extra time it would lead to fighting . . . more problematic behaviour . . . so it became trimmed a little bit. So they still have enough time to run around and eat, but they're not getting into so much trouble. I guess if they have less time on their hands to get into difficulties, then it's going to happen on fewer occasions.
>
> – English Teacher

Due to the behavioural issues and number of pupils during lunch breaks, it was decided that staggering the lunch breaks and reducing the duration of the break was for the best, particularly for minimising the number of incidents during this period. Although there may have been behavioural difficulties at the school in the past, the strategic decision to shorten the time and create a staggered lunch break helped to minimise disruptive behaviour during lunchtime, whilst arguably also training pupils to conform. This is about the students becoming agents of their own conformity and the school is exercising its strategic power relations in this case (Saldana, 2013). By this I am highlighting the power relations of the restaurant and the influence it can have. However, a parent expressed their concern about the impact of the staggered lunch break. For example:

> When my grandson was in Year 11, they used to be last because of the staggered lunch and he used to say that there's never anything left [in terms of food] . . . what was left, was dry.
>
> – Intervention TA, Parent to Pupil

Based on her grandson's experience, the intervention TA identified an issue with regards to the lack of food left over as a consequence of this staggered lunch system. Julia highlighted how this system meant the quality of food was being affected so the impact of this staggered system extended beyond just a lack of time in terms of a quality lunch break.

The English teacher suggested that less time during lunch led to a reduction in behavioural difficulties arising, but there was no evidence in my data to suggest that this had reduced issues of truancy. The restaurant is run in a particular way and in order to better understand how it is operated, reference has been made to the field notes to capture an accurate measure of the logistics of how it all works. For example:

> younger year groups line up by the smaller buffet line, waiting patiently to be seated. The teacher then allows 6 pupils with packed lunches to make their way to the tables, whilst others wait. There are two single files of pupils, one with packed lunches and others waiting to make their way to a parallel queuing system.
>
> – Field Notes

There are two queues of similar length running parallel to one another around the outside edge of the school restaurant with all the tables in between. So this means that those with packed lunches are potentially allowed more time to eat compared with those waiting to be served. With a reduced lunch break this could have a negative impact on pupils' school meal experiences. This system is interesting as I am yet to come across this anywhere else and at any other school in which I have taught, visited or attended as a pupil.

With this staggered lunch, one view is that it could help pupils to learn from their peers in different year groups to mix together. Although, the evidence suggests that younger pupils were separated from older pupils. For example:

> There is a clear separation between the younger and older pupils, even though the tables and chairs for the older pupils, who are yet to arrive are free. Chairs are a little smaller for the younger pupils, who take up their seats in the usual place.
>
> – Field Notes

From the preceding field note, pupils are separated by year groups in the school restaurant. This is a missed opportunity for the school who could make more of having multiple year groups in at the same time and not separating them in terms of their seating positions. Although the restaurant was designed to allow everyone to mix together, the staggered lunch system created pressure preventing this from happening, meaning that an opportunity for learning from older pupils was lost. This brings us to questioning the effectiveness of this staggered lunch system. Overall, it seems that the school made a strategic decision to stagger the lunch break in order to ensure all pupils were able to consume their lunches with little behavioural disruption. There has been a growing trend in surveillance studies and the study by Pike and Leahy (2012) is one example that recognises the disparity between school food and eating. Pike and Leahy (2012) explore school food pedagogies, more specifically how school food choices are regulated, which is one form of surveillance. The work of Pike and Leahy (2012) links closely to my research in terms of surveillance of school food but I am referring specifically to the lack of time as one form of regulation imposed by Peartree Academy. In relation to my research, this could arguably hinder pupils' opportunities in terms of having enough time to eat and converse with one another. It would appear that the original aims of the restaurant to create a community forum is being diverted by the need for surveillance.

PUPIL PARTICIPATION

Pupils are the main participants in the restaurant. This part draws on the negotiations which take place between pupils who conform and those who choose not to. It is useful to note that whilst conformity is central to this debate, it is said to take many forms in small groups in society and it may be the result of direct and overt explicit pressure, even though this may be unconscious (Saldana, 2013). This form of surveillance has a knock-on effect on pupil's behaviours to conform. The following field note illustrates how this is happening, and suggests that there are two things going on; one pupil is conforming and participating whilst the other is resisting and encouraging the one who is conforming to resist. For example:

> *The catering staff and other staff around the restaurant signal 2 minutes for two pupils who have been sitting at the table a while, still eating.*
> *1.08 pm*
> *As the older pupils prepare to enter the restaurant, these two pupils begin finishing their food.*

CATERING STAFF 1 – 'Come on guys, you have a couple of minutes.'
STAFF 1 – 'Guys, you came in the same time as us, there is no excuse.'
PUPIL 1 – 'Ok.'
PUPIL 2 – 'Why are they making us rush?'
PUPIL 1 – 'I'm going to leave this food.'
PUPIL 2 – 'No, don't you dare, they can't do that, that's child abuse. I am finishing my dinner, my pudding how I want.'
STAFF MEMBER – 'Right, let's get moving. Let's go, Lana, you've had enough time now, it's ten past 1!'

[The older pupils begin to make their way into the restaurant as they discuss their school day.]
[Pupils from Years 7 and 8 who should have another five minutes to finish as their lunch finishes at 1.15 pm.]

– Field Notes

The pupils then had to leave without finishing their food. There are different ways in which pupils operate in the restaurant, some are co-operative and eat quickly and leave the restaurant whilst others do not like to be rushed. Although the preceding example highlights how resistance was worn down, it was conformity that won. Again, this leads back to the point regarding the restaurant as a surveillance mechanism. The restaurant has been created to allow for social interactions to take place away from the pressures of the classroom (School Food Trust, 2007) and the social benefit here is that pupils want to have nice conversations and staff should be encouraging them to do this. Instead, the staff member has missed the opportunity of allowing pupils to take their time which means that they are being moved along in a conveyor belt fashion. Therefore, another structure is being superimposed on the restaurant, which goes against the aim of the school restaurant, which is to foster opportunities for social learning to take place and help to create a sense of community.

Some of the pupils are being pushed out of the restaurant to make way for other year groups. As discussed by Wills et al. (2015: 82), for schools with dual purpose dining halls (i.e. used for dining as well as teaching), the times of using the dining hall for eating became restricted. Consequently, this had a negative impact, meaning time for eating was significantly reduced. If this is happening regularly, then pupils have even less time for eating and having opportunities to converse with one another. For example:

Catering staff are wiping tables down in preparation for the next lot of pupils to arrive. Staff and pupils around the restaurant are tucking in chairs to maintain a level of cleanliness to the appearance of the restaurant.

– Field Notes

From the preceding example, the restaurant becomes redundant as a space for fostering opportunities for social interaction. Another pressure which works against the restaurant is that pupils are discouraged to eat lunch anywhere other than in the restaurant and they are not allowed to leave the premises. However, whilst most conform to this regulation others interfere in the process and try to find a way out of the school. It is said that schoolchildren are placed in an institution with strict rules and regulations that they are meant to conform to, but in reality they rarely do (Danaher et al., 2002). For example, the principal described how whilst most pupils conform and participate in the restaurant, there are some pupils who often resist the school lunch period by trying to climb the fences. For example:

> We don't allow our pupils out for lunch, although we have a few who climb the fences.
>
> – Principal

In a conversation with the principal, it was noted that truancy had been an issue and one regulation was that children should not eat anywhere outside of the restaurant. This was an interesting observation and although I did not see pupils climbing the fence, I heard this directly from the principal. It could be argued that the environment is being controlled as pupils are encouraged to eat in the restaurant and eating elsewhere is frowned upon.

Furthermore, in terms of resistance to the rules set by staff, some pupils who choose to stay in the restaurant still find ways of resisting and one pupil described how this was happening. For example:

> They get a baguette yeah, and they get away with this, they wrap it up in tissue, put it in their bag, go outside and eat.
>
> – Year 7 Pupil

Whilst some pupils seem to obey the rules set by staff, others seem to avoid this space. In this example, the pupil describes how pupils are eating outside and maybe this could be something to do with a lack of time or wanting to spend more time playing outside during the lunch break period. However, the critique of increasing surveillance places a focus on the presumed changes it might cause in space and social practices (Koskela, 2003: 294). As a consequence, surveillance could lead to a vicious circle of defence and resistance (Koskela, 2003).

Overall, pupils are being treated as if they are on a conveyor belt compared with a normal restaurant where people have more time to eat and converse. The rules set by staff work against the restaurant being a space that will enable pupils to develop opportunities for social learning. Surveillance techniques are being used in order to monitor where pupils eat but this monitoring means staff are focusing their energies on managing the behaviour of pupils, instead of allowing them to participate in a space which was originally designed for eating and social learning.

STRUCTURE AND CONTROL

The arrangement of the school restaurant is closely monitored by staff. The structure of the restaurant is a fragile one as it exposes both pupils and staff. It is fragile because staff seem to be patrolling pupils and this goes against the aims of the restaurant, which involve allowing pupils to mix and converse with each other. By structure, I am referring to the rules and regulations imposed on pupils by staff in the restaurant. The restaurant was designed to allow social interaction to take place and to make the eating experience a pleasant one. However, my evidence indicates that pupils are being controlled by staff, led by the leadership team. Visibility in the form of physical presence is of importance here as the senior leadership team are making a point of ensuring their presence is known to pupils in the restaurant. The leadership team and the principal are able to exercise power through visibility and invisibility, meaning they place themselves under the gaze whilst also removing themselves from it (Niesche, 2011). I am referring to the presence in which the leadership team and principal are in the school restaurant but also that power can be exercised through both visibility and invisibility. In my research, the principal or leadership team can enter the school restaurant at any time. This means they are able to exercise power even if they are not physically present in the restaurant.

Firstly, it was interesting to hear the perspective of the deputy principal who highlighted how he line managed the restaurant manager. For example:

> Yes, I line manage the restaurant manager . . . it's how I run the school, so I'm down there every day . . . I always, always, always have lunch in the restaurant, I always, always have informal meetings in the restaurant.
>
> – Deputy Principal Finance and Resources

The deputy principal emphasises how he always makes time to eat and have informal meetings in the restaurant. The deputy principal is responsible for line managing the restaurant and this means he has control of what goes on in the restaurant on a daily basis. This was also observed in my field notes. For example:

> There definitely seems to be a strong level of behavioural management around the restaurant. A very well staffed area with lots of presence. Also, the busiest time of the school day.
>
> – Field Notes

I could see that the restaurant was a space carefully monitored by staff and the leadership team, who were visible in the restaurant. Usually, most managers appear at the beginning and end of a period or day and are generally not a seen presence as they delegate tasks. However, this manager is not like that; he has presence in the school restaurant. Secondly, the positioning of the deputy principals in and around the restaurant signal how behaviour management is a high priority.

I noticed and observed how staff chose where they sat in the restaurant. The senior management team were consistently positioned at the circular tables, which meant they were in a position of visibility and were able to have presence in the restaurant. Other staff chose where they seated themselves in the restaurant. For example:

> The senior staff members including the Principal and Deputy Principals entered the restaurant at different times during the lunch period, although there were occasions when they were all present. They ate together on one of the circular tables. This table was located in the middle of the restaurant. It would mean that they would have presence in the centre of the restaurant and be able to see the school entrance [this was arguably a way of monitoring everyone in the restaurant who might enter the school and also ensuring presence is maintained].
> – Field Notes

It is through this visibility and use of power that staff are able to get pupils to conform to the rules and regulations of the restaurant. Furthermore, staff are subject to this internalisation of control, meaning arguably they are also conforming. As the restaurant is designed to expose everyone from a visual point of view, this means the senior staff team are also able to watch staff in the school. Therefore, the restaurant becomes a space which offers a view of the school entrance. However, there was not any evidence available to support the idea of staff being watched. This is the most recognised panoptic principle; the basic nature of the exercise of disciplinary power involves regulation through visibility (Hannah, 1997: 171).

The deputy principals were seated on one of the few circular tables, situated in the middle of the restaurant. They chose to sit in this particular part of the restaurant regularly and pupils would rarely sit at these tables, knowing that this table was usually used by staff. However, there were occasions where perhaps a couple of pupils would join the staff. For example:

> I noticed how often the Deputy Principals would be facing the reception area in order to have a clear view of who comes in and out of the school. From time to time, some pupils would be sat with the staff on

these tables, but this was rare. Staff would frequently walk into and across the restaurant as it was an unavoidable space in the school.

– Field Notes

The deputy principals were able to see the main entrance into the school from this position. Whilst they had their back turned to pupils, their presence was paramount. The deputy principals occasionally walked through the restaurant throughout the school day as well as during the lunch period. This allowed staff to keep a watchful eye on those who entered the school as they passed through the restaurant on their way to the classroom.

Having developed a more structured set of field notes, I was able to record how control was maintained during other periods of the school day. The restaurant was utilised from the start of the day to the very end. I developed a chart to help record and reflect on certain situations (see Table 4.2).

Table 4.2 Illustrating observations

Period	Year group	What I observed	What I heard	Further thoughts
Post-breakfast	Younger year pupils	A group of teachers evenly spaced walking in a single file with the younger year groups, walking out of the school for an excursion.	General pupil chatter of excitement.	A **structured** and **controlled** environment, which I was able to see from the restaurant.
Post-lunch break time	Year 7/8	One pupil zooms past the restaurant. There is one member of staff in the area, who is supervising, a lot less than the usual numbers of staffing during the post-lunch break time.	Pupils seem very lively. Generally comments included – 'Where you going?' Staff member wearing blue hoody – 'Ladies, ladies, ladies, ladies, right, let's get together and start making our way to the next lesson.'	Pupils are listening well regardless of the loud environment during break time – very **controlled**.

The field notes in Table 4.2 reveal the level of structure and control that is shaping the day-to-day running of the restaurant space. The dialogue in this table highlights how staff are monitoring the movements of pupils. This was also noted in my field notes, particularly during the lunch break where pupils were being controlled. For example:

Lunch break
There definitely seems to be a strong level of behavioural management around the restaurant. A structured and well-staffed area of the school. Also, the busiest time of the school day.
12.30 pm
As most pupils are seated, the restaurant seems to be fairly controlled, with only a few pupils wandering around with trays, back and forth.

– Field Notes

The school lunch is known as being one of the busiest periods of the school day and good supervision during this unstructured time is essential (Thorpe, 2004). From my field notes, it is evident that the staff at Peartree Academy have been prepared for this time of day.

It was quite early into my observations that I noticed how the restaurant was being controlled or at least, there were lots of staff around including teachers, who were patrolling during the lunch break. This was unusual compared to my experience of working in a school or being a pupil. One way in which this is perhaps beneficial is that staff are able to keep a close eye on pupils who enter the school late.

The evidence in the following field note describes how the deputy principal welcomes pupils who arrive late to the school. The usual experience of this would be that a pupil who arrives late signs into reception and provides evidence for their lateness. The unusual experience was that the deputy principal, Adam Walker, intervened to ensure pupils who arrived late were addressed. It was unusual to see a member of the leadership team working in this way. For example:

Latecomers make their way, through towards the main corridor area, as the Deputy Principal (Adam Walker) welcomes them, taking a friendly approach in conversing with them whilst walking along the main corridor, through the restaurant.

– Field Notes

For me this seemed to be a way of ensuring someone is able to watch those who arrived late on a regular basis. Adam Walker regularly picked up the latecomers, walking along with them as they entered the school in the direction of the main corridor, leading to all the classrooms. This is about the leadership team using visibility in the restaurant area to control pupils.

Therefore, the senior leadership team are engaging with pupils in managing their behaviour in the restaurant. Two examples of how this is happening include firstly, pupils arriving late to school, which include those who are unable to avoid passing through the restaurant on their way to class; and secondly, the ways in which pupils are behaving in the restaurant during the lunch break. These are two instances in which the leadership team seem to be present. The lateness of pupils is just one way in which monitoring of pupil behaviour in and around the restaurant is taking place. It draws specifically on how staff are regulating behaviour in the restaurant. In this instance, the senior staff member is speaking to a colleague whilst addressing a pupil for lateness as he makes his way through to the school restaurant towards the classrooms. For example:

One of the deputy principals sits and chats with a colleague whilst addressing pupils as they walk around the restaurant towards their lesson (latecomers).

SENIOR STAFF MEMBER – 'Have you been to lesson yet? You're late!'
PUPIL – 'Yeah.'
SENIOR STAFF MEMBER – 'What lesson?'
PUPIL – 'Enterprise.'
SENIOR STAFF MEMBER – 'Where's your tie?'
PUPIL – 'In my pocket, sir.'
SENIOR STAFF MEMBER – 'Good morning, by the way.'
PUPIL – 'See you later.'

– Field Notes

This example highlights how behaviour is monitored by staff. There are two things going on here. Firstly, the pupil is being quizzed regarding his lateness and secondly, the staff member is trying to get the pupil to conform by asking him about his tie. Usually pupils who are late go straight to class in my experience. However, at Peartree Academy, pupils are greeted by senior members of the staff team. In this situation, the staff member is paying lip service to a particular regulation as we can assume the pupil will probably not wear the tie. The way in which the staff member interferes is different to how teachers would perhaps manage pupils in a classroom. The fragility here is with the use of the staff member's authority and learning to use it in a way which allows the pupil to respond to it. So, it is a question of whether this pupil will actually listen or choose not to and it has already been said that pupils are not always inclined to conform (Danaher et al., 2002). The fragility then lies in the balance between monitoring and managing behaviour.

Overall, this part highlighted how the restaurant is managed and provided a discussion on the complexities of the staggered lunch break. Whilst priority is given to ensuring all pupils consume their lunches, it can be argued that these pressures against the restaurant mean that once pupils have finished their lunches they are being asked to move on. Consequently, this defeats the purpose of why the restaurant was designed in the way that it was. It was to ensure everyone had enough time to converse whilst eating. Clearly, the restaurant is a fragile space in the way in which negotiations between pupils and staff take place and also about how they are using it. Therefore, it is how these social relationships are being constructed and reconstructed or even not being allowed to develop (Punch et al., 2013). Part two of the chapter introduces a discussion on the modelling of eating behaviours in the restaurant.

Modelling eating behaviours in the restaurant

This final part of the chapter highlights the impact of the restaurant on how behaviour is being modelled and draws on the notion of the restaurant as a community forum. As introduced earlier in this chapter, it is the definition of modelling by Eliassen (2011) that is being used. He argues that teachers, like parents, have an influence and are able to model appropriate eating behaviours. In terms of modelling then, the discussions highlight how pupils are inducted and trained in terms of how to behave in the school restaurant. Furthermore, Osowski et al. (2013) identified how the pedagogic meal was practiced in Swedish dining halls focusing specifically on the teacher's interaction with the children and the roles being modelled by teachers.

This part is made up of three segments, which include, 1) modelling behaviour: staff roles, 2) modelling behaviour: staff member or temporary parent? and 3) modelling behaviour: the school restaurant as a third teacher. The first part discusses how certain roles are being presented by staff which carry mixed messages. The second part demonstrates how the staff role is extended beyond their role as staff members. The final part highlights the symbolic messages being carried by the school restaurant as a space.

MODELLING BEHAVIOUR: STAFF ROLES

The role of staff during mealtimes is said to be an influential part of children's mealtime experiences (Sepp et al., 2006). During the observations carried out, I noticed how some staff conducted themselves in terms of their roles in the restaurant. Although observations were carried out throughout the school day, the most notable occurrences took place during the lunch break. Therefore, this section highlights the types of staff roles on display and how they interacted with pupils in the restaurant and this is presented in a table in my field notes (see Table 4.3). The following example from my field notes describes how staff presented certain behaviours in a strategic way. This has been represented in the form of a chart (Table 4.3), which was developed during my data analysis. For example:

> I am beginning to notice the younger pupil's supervisors being more animated, particularly 1 male member of staff, who is using body language to make the suggestion of how food will make you stronger! [At this point, I am noticing multiple interactions between staff and pupils in the restaurant and I decided to draw up a list of the different type of strategies being implemented by staff in the restaurant.]
>
> – Field Notes

There are two things happening here: firstly some of the staff are encouraging pupils to participate in the restaurant by modelling certain behaviours

Table 4.3 Illustrating staff roles

Staff role/strategy	What I observed
Promotor of healthy eating benefits	Mr Muscle [A member of staff, is very animated and working with the younger pupils, suggesting eating all your food will allow you to gain physical strength].
Behaviour support	'Calm down' [The staff member spread his arms across his body wide in order to establish control of the situation or at least the attention of pupils. He then slowly lowered his arms signalling that they needed to calm down].
Managing the trays and pupil movement	'Let me take those trays for you' [This was the staff member who wanted to ensure any disruptions were kept to a minimum in the restaurant].
Sociable member of staff	'How you getting on?' [This was the member of staff who would mingle with pupils and have regular chats. For instance, the staff member would sit with pupils and move around every so often and pick out different topics which included football, music and technology].

(Continued)

Table 4.3 (Continued)

Staff role/strategy	What I observed
Listener	'How are things at home?' [This member of staff would seek out pupils who are perhaps isolated in the restaurant and are seated alone. The focus seems to be on monitoring whilst listening].
Observational member of staff	Reserved [This was the staff member who regularly took a step back to look over how pupils were actually behaving in the restaurant].

whilst others seemed to be focused on monitoring the restaurant and are interested in surveillance. The first staff role was the 'promotor of healthy eating' who used body language to express his thoughts e.g. 'fist pumping' to acknowledge a piece of chicken on a plate, signalling that this will make you stronger. For Garvis and Pendergast (2017) there is much to be said about the role of food, which helps boost knowledge on food choices and these can have a positive influence, particularly for children. This signals that pupils can develop their physical health through healthy eating practices through staff modelling in which healthy eating choices can be made. This is about inducting and training pupils about what to eat. Furthermore, as previously discussed, Osowski et al. (2013) also identified different types of teachers in the dining hall and the three types as discussed in chapter two included 1) the social teacher, 2) the educational teacher and 3) the evasive teacher. They found a positive association between these as pupils had some role models to look up to, particularly the social and educational teachers.

The second staff role that I identified was the 'behaviour support strategy' which involved spending time using body language to enable pupils to learn how to conduct themselves. The role of 'managing the trays and pupil movement' involved the staff member ensuring that everything is managed appropriately to avoid any disruption, and this includes collecting the trays on behalf of pupils and minimising pupil movement. The 'sociable' member of staff created a social occasion during the school lunch period, which involved a high level of social interaction with pupils. This involved conversing with pupils regularly. The 'Listener' would observe and pay attention to the pupils' wellbeing, both in and outside of school whilst monitoring the movements of pupils. The 'observational' member of staff took a step back and watched the pupils throughout the restaurant carefully, noting any misbehaviour, whilst discreetly challenging pupils where necessary. The list of roles in Table 4.3 present an ambiguity in what is happening in the school restaurant. Some of the staff are taking the opportunities to develop social learning in pupils (sociable, listener) whilst others seem to be more concerned with monitoring and inducting pupils' eating behaviours (promotor of healthy eating benefits, behaviour support, managing the trays, observational).

The senior behaviour teacher described how pupils were given opportunities to interact in a social way and that staff seem to spend a considerable amount of time in the restaurant with the pupils. For example:

> I think pupils get to interact in a social way, the staff are very good you may have seen them eating with the pupils, the staff then have a responsibility to model that social behaviour.
>
> – Senior Behaviour Leader

As the staff are eating with the pupils, they are able to tap into the opportunities to promote positive eating behaviours. The idea here is that staff are responsible for inducting and training pupils in how to conduct themselves in the restaurant. For example:

> Staff are generally seated on the circular tables, which are in the middle of the restaurant and some of them are also occupied by pupils so staff and pupils are eating together.
>
> – Field Notes

The restaurant in this view below potentially allows pupils to observe how to behave. Year 7 teacher Rachel Jones discussed how pupils are able to observe staff on duty and identifies this as a form of interaction. For example:

> in break and lunch times specifically, I think in the restaurant, there's people on duty so there's promotion there, of how to interact in a free environment, a lot of our students struggle with unstructured times, so there's staff there to reassign people and redirect people that are not coping in that unstructured time.
>
> – Year 7 Nurture Teacher

The Year 7 nurture teacher talks about the promotion of positive behaviours where pupils seem to struggle with unstructured times. For Rachel, this was about staff making a judgement about which pupils needed the support. Overall, this part highlights some of the strategies used in supporting the original aim of the restaurant, which was to create opportunities for social learning. However, there seem to be some pressures working against this as staff seem to be more concerned with monitoring pupils. Whilst the intentions of some of the staff were positive, others appeared to be more concerned with monitoring pupil behaviour.

MODELLING BEHAVIOUR: STAFF MEMBER OR TEMPORARY PARENT?

Parents play an important role in shaping children's attitudes and values about food by acting as role models (Richards and Smith, 2007). Staff at Peartree Academy are taking on the responsibility, typically attached to the role of a parent, just like staff in many other schools. However, the staff role in this school is different, particularly during the breakfast and lunch period where staff are taking on the role of ensuring that pupils also make good food choices (Birch, 1980). This section is about the modelling of food choices and the role of staff as temporary parents. For Eliassen (2011) parents influence children's eating behaviours and attitudes towards food through observation at home. For instance, children are seen to consume the same or similar food to their parents (Eliassen, 2011). The following field note demonstrates how staff in the school used the restaurant to 'mother' some pupils. For example:

> So one thing I noticed was how the exams officer and administration assistants spent time speaking to pupils about how they were getting on and how their day at school had been going. It felt like they were being mothered. It wasn't just the younger pupils; it was the older pupils too. I guess when pupils come away from the home environment, they still need to be supported or at least someone needs to replace the role of the parent for them to feel at home or at least comfortable in the surroundings [it is interesting to see how non-teaching staff involved in administrative roles also took time out to communicate with pupils].
>
> – Field Notes

Eliassen (2011) found that the role models had an impact on the children's subsequent food choices, with the exception of the adult. It was found that children were more likely to experiment with unfamiliar foods upon observing a role model. Staff at Peartree Academy also have responsibility for ensuring that pupils are equipped with the right attitudes in making good food choices and conducting themselves appropriately. For example:

> think we've got a lot of challenges with parents. So we're trying to get the children to teach the parents, because it is quite shocking when the children come in and eat sweets and high sugary things in the morning.
>
> – Assistant Principal Phase 1

The assistant principal describes how staff intervene by trying to encourage pupils to eat healthier foods, which highlights their role as a temporary parent to pupils. Lucy observes how pupils come into school with sugary sweets in the morning and this becomes about educating children on making better food choices. It is also about taking that home and allowing pupils to teach their parents. In this case, I would argue that eating behaviours of pupils can be influenced at Peartree Academy through staff modelling and taking the role of a parent in the school restaurant (Eliassen, 2011). For example, Klesges et al. (1991) identified how children selected different foods when they were being watched by their parents compared with when they were not, and overall the children made better food choices when they were being watched. So, this is about the staff member taking on the position of parent in the restaurant. In another study conducted by Birch (1980) on children's food preferences, peer modelling saw changes in their preferences for vegetables. Therefore, there does seem to be a link between inducting and training pupils about eating behaviours.

Overall this part argued how the school restaurant can be used to foster opportunities for social learning, yet the evidence suggests that, in this school at least, there is not the time and space to allow this to happen. The examples from this section highlight how positive eating behaviours are being modelled and encouraged. Therefore, it can be argued that the restaurant is playing a key role in allowing for modelling to take place.

MODELLING BEHAVIOUR: THE SCHOOL RESTAURANT AS A THIRD TEACHER

It is common to assert that certain areas of the school, including empty spaces and waiting spaces outside, are often overlooked, and this introduces the notion of the 'third teacher' (Nicholson, 2005). Nicholson (2005) argues that pupils have an awareness of the symbolic messages attached to a school building and points out how schools are now becoming more attractive in their appearances in order to foster opportunities for social learning. Furthermore, the notion of the 'cam-era' as highlighted by Koskela (2003) also supports the ideas put forward by Danaher et al. (2002) regarding self-regulation. This part highlights the impact of the school restaurant as a fixture in the school and how it can have a positive influence on pupils' eating behaviours. It is argued that the restaurant is one part of the school building which acts as a third teacher in order to foster opportunities for social learning.

In the following field note, I observed how the restaurant potentially acted as a third teacher (Nicholson, 2005). By this, Nicholson (2005) highlighted the importance of the environment in complementing the educational and social support of the pedagogy. In this example the restaurant is being described as a common room space:

It is also a space in the school which I can see has the potential to allow pupils and staff to interact with one another. So far, outside of the lunch period, I have seen the restaurant in constant use, particularly at times when food isn't being consumed. It is what I would call a common room space, where pupils and staff are able to sit, talk and catch up throughout the school day. The main activity at times outside the lunch period is purely conversation between various year groups, staff and parents.

– Field Notes

This highlights how the restaurant can be used to help foster opportunities for conversation and social learning to take place. If it is used well then this could be achieved. The wall displays in the school restaurant carry messages which are directed at pupils. I was able to capture the content of the display by recording them in the field notes. Eight key messages were displayed above the buffet aisle and kitchen. This was in view of all of those who entered the school restaurant. For example:

1   '5 a day' – Important to eat five portions of fruit and vegetables a day.
2   'Mealtime' – I'm eating three meals a day including a healthy school lunch.
3   'Me-size meals' – I'm eating meals that are the right size for my age, not as big as grown-ups.
4   'Cut back fat' – My family are changing how we cook to make our meals healthier.
5   'Snack check' – Lots of snacks are full of fat, sugar and salt so I'm eating healthy snacks!
6   'Sugar swaps' – I'm swapping sugary drinks for water, milk or unsweetened fruit juice.
7   'Up and about' – After I've been sitting still for a while, I'm jumping up and doing something more active.
8   '60 interactive minutes' – I'm spending at least 60 minutes walking, playing sport, running around, or playing outside every day.

– Field Notes

For me, this is a way that the school was able to attempt to control the daily eating behaviours of pupils in the school restaurant, if they are to read these messages. These displays are arguably an example of how visual aids were used in the school to help foster opportunities for developing positive eating behaviours (Eliassen, 2011).

Following several observations, when one of the staff members mentioned that there was a camera in the restaurant, I also noticed it. For example:

> I have also noticed a camera in the restaurant, one which captures the whole school restaurant and reception area as well as the main school corridor.
>
> – Field Notes

It was in fact a legitimate camera as opposed to an imitation. One pupil described her view on how she felt about the school camera in the restaurant. For example:

> like they have cameras, there's this massive security camera, some people do you know what they do yeah.
>
> – Year 7 Pupil

This leads me back to the notion of surveillance. This is described by Koskela (2003) as 'the cam-era' – an era of endless representations (Koskela, 2003: 292). His study on the contemporary urban panopticon also looked at 'space' as a crucial factor in explaining social power relations.

Consequently, part of our socialisation influences us to make ourselves the subject of our own gaze, meaning that we are constantly monitoring our behaviours (Danaher et al., 2002: 54). Therefore, it can be argued that both staff and pupils become subject to surveillance of this type. Through surveillance cameras the panoptic technology of power is electronically extended (Koskela, 2003).

In terms of this notion of the third teacher, it could be said that the restaurant was designed to allow pupils to converse with one another whilst gaining opportunities for learning. However, other staff are more concerned with managing behaviour in the restaurant and it is these pressures which seem to be working against the restaurant being an informal community forum. The restaurant has presence in the school and clearly has the potential to foster opportunities for social learning which are being taken by some of the staff in the school. This part has demonstrated how a dining space in a school can be used to enrich the experiences of pupils.

The 'restaurant' as the 'third teacher' also has an inbuilt ambiguity in the way the tables are set out and the camera on the wall. The 'third teacher' has the same dual role as the staff. In this case, whilst an attempt has been made to create a community forum, it is one which is also being monitored.

The argument presented in this chapter highlighted two narratives running in parallel, firstly that the restaurant allowed pupils to converse and interact and secondly that the pressures of the lunch break timings and the ways in which staff patrolled the restaurant worked against this. Just like many other schools, maintaining behaviour was highly prioritised by staff compared with trying to influence pupils' eating behaviours during the school lunch period. There is a level of monitoring and surveillance taking place and some of the staff are too focused on managing pupils, although other staff are taking these opportunities of modelling positive eating behaviours. The tension between providing social learning opportunities and surveillance carried out by staff presents limits, meaning there are limits on the pupils being able to develop eating behaviours. Whilst there is some evidence to suggest that social learning could potentially take place in the restaurant, the pressures against this are preventing this from happening.

Part one of this chapter considered some key discussions involved in the running of the school restaurant, particularly focusing on the staggered lunch, pupil participation and staff presence. There was some evidence to suggest that pupils were resisting participating during the school lunch period, but further investigations which concentrate on this idea need to be carried out in order to develop a stronger argument.

Part two explored staff influence on the school restaurant, whilst also drawing on the notion of the school 'restaurant' as a 'third teacher'. It high-lighted how power can be interpreted in a more positive way; to ensure pupils have access to role models in order to develop eating habits and behaviours. The restaurant in this view acted as a platform for inducting and training pupils, meaning positive eating behaviours are being filtered through the behaviour of staff and other pupils. Overall, the idea here is that staff have the opportunity to play an active role in training and inducting pupils in order to prepare them for developing their social skills.

Conclusion

This chapter argues how opportunities for social learning are being missed due to the lunch timing constraints, rules and regulations imposed by staff on pupils. Pupils are being treated as if they are on a conveyor belt compared to a normal restaurant, where people have more time to sit, eat and converse. The rules and regulations imposed by staff are preventing them from modelling those positive eating behaviours for pupils. Therefore, in this case, there is more evidence to suggest that the restaurant is a surveil-lance mechanism as opposed to a community forum. However, from the school's perspective, when trying to create a community forum, these are the difficulties that those who are leading the school are facing.

The school originally set out with good intentions to build a restaurant which modelled an internal version of a community forum for all. It is the pressures of the rules and regulations that work against the school being able to achieve this. Unfortunately, this means the school restaurant becomes subject to a complex debate and one which is fragile, and ultimately the evidence suggests that opportunities for developing social learning are missed. It would be useful to carry out a further investigation into school dining halls, in order to investigate further lines of inquiry which continue developing key debates whilst trying to establish some recommendations for other schools to adopt in the future.

References

Birch, L. L. (1980) 'Effects of peer models' food choices and eating behaviours on pre-schoolers' food preferences', *Child Development*, 51, pp. 489–496.

Calvert, M. and Henderson, J. (1998) *Managing Pastoral Care*. London: Continuum.

Danaher, G., Schirato, T. and Webb, J. (2002) *Understanding Foucault*. New Delhi: Motilal Benarsidass.

Devi, A., Surender, R. and Rayner M. (2010) Improving the food environment in UK schools: Policy opportunities and challenges, *Journal of Public Health Policy*, 31 (2), pp. 212–226, Available at: http://www.jstor.org.ezproxy4.lib.le.ac.uk/stable/pdfplus/40802312.pdf?acceptTC=true&jpdConfirm=true (Accessed: 13 August 2019).

Eliassen, R. K. (2011) 'The impact of teachers and families on young children's eating behaviours'. Available at: www.naeyc.org/files/naeyc/Eliassen_0.pdf (accessed: 10 February 2019).

Flowers, R. and Swan, E. (2015) *Food Pedagogies: Critical Food Studies*. Surrey: Ashgate Publishing.

Foucault, M. (1980) 'The eye of power', in Gordon, C. (ed.), *Power/Knowledge: Selected Interviews and Other Writings 1972–1977 by Michel Foucault*. Sussex: Harvester Press.

Garvis, S. and Pendergast, D. (2017) *Health and Wellbeing in Childhood*. Cambridge: Cambridge University Press.

Hannah, M. (1997) 'Space and the structuring of disciplinary power: An interpretive review', *Geografiska Annaler*, 79B, pp. 171–180.

Klesges, R. C., Stein, R. J., Eck, L. H., Isbell, T. R. and Klesges, L. M. (1991) Parental influence on food selection in young children and its relationships to childhood obesity, *American Journal of Clinical Nutrition*, 53 (4), pp. 859–864.

Koskela, H. (2003) *Surveillance and Society*, 1 (3), pp. 292–313. Available at: www.surveillance-and-society.org/articles1(3)/camera.pdf (accessed: 8 June 2019).

Lalli, G. (2019) 'School mealtime and social learning in England', *Cambridge Journal of Education*. DOI: www.tandfonline.com/eprint/fHGEr7k8ebj55DVBY6YN/full?target=10.1080/0305764X.2019.1630367.

Lightfoot, L. (2007) 'Where's my lunch?' *The Guardian*. Available at: https://www.theguardian.com/education/2007/nov/13/schools.uk2 (Accessed: 23 August 2019).

Nicholson, E. (2005) 'The school building as third teacher', in Dudek, M. (ed.), *Children's Spaces*. London: Architectural Press.

Niesche, R. (2011) *Foucault and Educational Leadership: Disciplining the Principal*. London: Routledge.

Osowski, C. P., Goranzon, H. and Fjellstrom, C. (2013) 'Teachers' interaction with children in the school meal situation: The example of pedagogic meals in Sweden', *Journal of Nutrition Education and Behaviour*, 45 (5), pp. 420–427.

Pike, J. (2010) *An Ethnographic Study of Lunchtime Experiences in Primary School Dining Rooms*. Ph.D. Thesis, University of Hull.

Pike, J. and Leahy, D. (2012) 'School food and pedagogies of parenting', *Australian Journal of Adult Learning*, 52 (3), pp. 435–459.

Punch, S., McIntosh, I. and Emond, R. (2013) *Children's Food Practices in Families and Institutions*. London: Routledge.

Richards, R. and Smith, C. (2007) 'Environmental, parental, and personal influences on food choice, access and overweight status among homeless children', *Social Science and Medicine*, 65 (8), pp. 1572–1583.

Saldana, J. (2013) 'Power and conformity in today's schools', *International Journal of Humanities and Social Science*, 3 (1), pp. 228–232.

School Food Trust. (2007) 'A fresh look at the school meal experience'. Available at: https://www.school-portal.co.uk/GroupDownloadFile.asp?GroupId=1007950 &ResourceId=3301370 (Accessed: 23 August 2019).

Sepp, H., Abrahamsson, L. and Fjellstrom, C. (2006) 'Pre-school staffs' attitudes toward foods in relation to the pedagogic meal', *International Journal of Consumer Studies*, 30 (2), pp. 224–232.

Thorpe, P. (2004) *Understanding Difficulties at Break Time and Lunchtime*. London: National Autistic Society.

Wills, W., Draper, A. and Gustafsson, U. (2015) *Food and public health: Contemporary issues and future directions*. London: Routledge.

Zandian, M., Ioakimdis, I., Bergstrom, J., Brodin, C., Leon, M., Shield, J. and Sodersten, P. (2012) 'Children eat their school lunch too quickly: an exploratory study of the effect on food intake' *BMC Public Health*, 12 (351), pp. 1–8.

5   'Food for thought'

Conclusion

Conclusion

School meals have become a site for surveillance in that they are not necessarily seen as being a space of learning, but of consumption in a culture which continues to see children as machines for productivity. However, pressing concerns on obesity, health, wellbeing and attainment suggest that eating could and should be considered in educating the whole child and not merely seen as a small part of the school day. For such an event to occur so frequently, the school lunch period needs more attention, not only through investment of space, but in terms of interaction between children and teachers in particular. In light of these discussions, it is important to return to the original ideas of this study and unpack research aims and how they are informed.

To take the main research question, I have come to learn that the food environment has a huge impact on social learning in terms of the way in which school dinners are organised and served. With regards to the type of impact the food environment in a school has depends on the staff in the school. By staff, I am referring to the leadership team, teaching staff and administrative employees. Social learning opportunities can be both created and harnessed by taking the opportunity to recognise the school meals as a time of day to learn. Moreover, the space in which the school meal is designed also has a positive influence on the attitudes towards school meals for pupils. Food is a powerful force in that it can bring different groups of pupils and staff together where there is the potential to exercise and develop social etiquette. For me, the main research question enabled me to provide an overview of the important role of food in schools in relation to social learning.

In terms of the first subsidiary research question, eating behaviours of staff and pupils play a key role in fostering opportunities for social learning.

More specifically, staff hold the power in modelling behaviour in the school dining area and for this reason, investigating staff behaviour was crucial in providing evidence for how pupils can be influenced. Furthermore, pupils also act as role models to younger pupils and are therefore in a position of power to influence one another in making the transition from one year group to the next as well as the life transition outside of the school.

The second subsidiary research question has had a huge influence on my desire to seek out how teaching staff are able to promote social learning opportunities within a food environment. Teachers have a massive impact on pupils with regards to their participation and presence in the school dining area and their actions and techniques play a part in helping pupils to develop social etiquette, which is an important life skill. Moreover, lunchtime supervisors are also in a position of power and have an impact on pupils' mealtime experiences. There was a difference between teaching and non-teaching staff in teaching social learning. For example, the lunchtime supervisors had a better understanding of social learning and also took the opportunities to foster them in pupils.

Recommendations for schools, teachers and policy makers

In terms of recommendations for schools, it can be argued that the school mealtime should be extended and for more communication to take place between teaching and non-teaching staff. I would like to recommend that schools invest in putting aside more time for consuming school lunches. This could be particularly useful for new schools who face the challenge of allowing pupils to have enough time to eat whilst being able to interact with one another. The restaurant is a space for fostering social interactions which can help the life experiences of pupils. It can also encourage pupils to develop their people skills, which become transferable skills throughout their life span. Furthermore, there are currently no national standards for the length of the school lunch period, and little is known about the relationship between the amount of time pupils have to eat and school food choice and consumption.

For teachers, I would like to recommend my thesis as one of many which place an importance of the school meal as a social event which can help forge new and maintain existing relationships, where pupils, staff and parents are able to come together to develop fruitful relationships. I would like to recommend that teachers take the opportunity of mingling and socialising with pupils during the school lunch period as this can be a fragile time of day for some pupils who could benefit from social interaction. Furthermore, I would also like to reach out to teachers who are particularly struggling with managing pupil behaviour in and around the school during the lunch period. So, future research endeavours may attempt to address the importance of school dining spaces as another site for surveillance or whether more could be done to prioritise mealtime.

Index

Printed in the United States
by Baker & Taylor Publisher Services